MICROWAVE PROCESSING OF MATERIALS

COMMITTEE ON MICROWAVE PROCESSING OF MATERIALS: AN EMERGING
INDUSTRIAL TECHNOLOGY
NATIONAL MATERIALS ADVISORY BOARD
COMMISSION ON ENGINEERING AND TECHNICAL SYSTEMS
NATIONAL RESEARCH COUNCIL

Publication NMAB-473
National Academy Press
Washington, D.C. 1994

NOTICE: The project that is the subject of this report was approved by the Governing Board of the National Research Council, whose members are drawn from the councils of the National Academy of Sciences, the National Academy of Engineering, and the Institute of Medicine. The members of the committee responsible for the report were chosen for their special competences and with regard for appropriate balance.

This report has been reviewed by a group other than the authors according to procedures approved by a Report Review Committee consisting of members of the National Academy of Sciences, the National Academy of Engineering, and the Institute of Medicine.

The National Academy of Sciences is a private, nonprofit, self-perpetuating society of distinguished scholars engaged in scientific and engineering research, dedicated to the furtherance of science and technology and to their use for the general welfare. Upon the authority of the charter granted to it by the Congress in 1863, the Academy has a mandate that requires it to advise the federal government on scientific and technical matters. Dr. Bruce M. Alberts is president of the National Academy of Sciences.

The National Academy of Engineering was established in 1964, under the charter of the National Academy of Sciences, as a parallel organization of outstanding engineers. It is autonomous in its administration and in the selection of its members, sharing with the National Academy of Sciences the responsibility for advising the federal government. The National Academy of Engineering also sponsors engineering programs aimed at meeting national needs, encourages education and research, and recognizes the superior achievements of engineers. Dr. Robert M. White is president of the National Academy of Engineering.

The Institute of Medicine was established in 1970 by the National Academy of Sciences to secure the services of eminent members of appropriate professions in the examination of policy matters pertaining to the health of the public. The Institute acts under the responsibility given to the National Academy of Sciences by its congressional charter to be an adviser to the federal government and, upon its own initiative, to identify issues of medical care, research, and education. Dr. Kenneth I. Shine is president of the Institute of Medicine.

The National Research Council was organized by the National Academy of Sciences in 1916 to associate the broad community of science and technology with the Academy's purposes of furthering knowledge and advising the federal government. Functioning in accordance with general policies determined by the Academy, the Council has become the principal operating agency of both the National Academy of Sciences and the National Academy of Engineering in providing services to the government, the public, and the scientific and engineering communities. The Council is administered jointly by both Academies and the Institute of Medicine. Dr. Bruce M. Alberts and Dr. Robert M. White are chairman and vice chairman, respectively, of the National Research Council.

This study by the National Materials Advisory Board was conducted under Contract No. MDA 972-92-C-0028 with the Department of Defense and the National Aeronautics and Space Administration.
Library of Congress Catalog Card Number 94-66560

International Standard Book Number 0-309-05027-8

Available in limited supply from:
National Materials Advisory Board
2101 Constitution Avenue, NW
HA-262
Washington, D.C. 20418
202-334-3505

Additional copies are available for sale from: National Academy Press 2101 Constitution Avenue, NW Box 285 Washington, D.C. 20055 800-624-6242 or 202-334-3313 (in the Washington Metropolitan Area)

B-272
Copyright 1994 by the National Academy of Sciences. All rights reserved.

Printed in the United States of America.

DEDICATION

This report is dedicated to the memory of Joe Pentecost. Joe served as a member of the National Materials Advisory Board from 1988 until his death in 1992. This study was initiated by his efforts, and he was to have been a member of the study committee. We miss him as a person, always gracious and optimistic; as a technical leader with a vision of the future; and as a worker who always did more than his fair share.

Our lives take their meaning from their interlacing with other lives, and when one life is ended those into which it was woven are also carried into darkness. Neither you nor 1, but only the hand of time, slow-moving, yet sure and steady, can lift that blanket of blackness. Adlai Stevenson

COMMITTEE ON MICROWAVE PROCESSING OF MATERIALS: AN EMERGING INDUSTRIAL TECHNOLOGY

DALE F. STEIN *Chairman*, Michigan Technological University, Houghton
RICHARD H. EDGAR, Amana Refrigeration, Inc., Amana, Iowa
MAGDY F. ISKANDER, University of Utah, Salt Lake City, Utah
D. LYNN JOHNSON, Northwestern University, Evanston, Illinois
SYLVIA M. JOHNSON, SRI International, Menlo Park, California
CHESTER G. LOB, Varian Associates, Inc., Palo Alto, California
JANE M. SHAW, IBM-T.J. Watson Research Center, Yorktown Heights, New York
WILLARD H. SUTTON, United Technologies Research Center, East Hartford, Connecticut
PING K. TIEN, AT&T Bell Laboratories, Holmdel, New Jersey

Government Liaison Representatives

WILLIAM COBLENZ, Advanced Research Projects Agency, Arlington, Virginia
ALAN DRAGOO, Department of Energy, Washington, D.C.
SUNIL DUTTA, National Aeronautics and Space Administration, Cleveland, Ohio
CHARLES LEE, Air Force Office of Scientific Research, Bolling Air Force Base, Washington, D.C.
WILLIAM MESSICK, Naval Surface Warfare Center, Silver Spring, Maryland
JOHN W. WALKIEWICZ, U.S. Bureau of Mines, Reno, Nevada
WALTER ZUKAS, Army Materials Technology Laboratory, Watertown, Massachusetts
National Materials Advisory Board Staff
    THOMAS E. MUNNS, Senior Program Officer
    AIDA C. NEEL, Senior Project Assistant

ACKNOWLEDGMENTS

The committee is most grateful to the many individuals who took the time to make very informative and useful presentations to the committee.

Speakers at the August 12, 1992 meeting included:

Dr. Mark Janney, Oak Ridge National Laboratory, *processing of ceramics and decontamination of concrete*

Dr. Joel Katz, Los Alamos National Laboratory, *processing of ceramics and hazardous waste processing*

Dr. Jes Asmussen, Michigan Sate University, *electrical engineering considerations related to microwave processing*

Dr. Martin Hawley, Michigan State University, *application of microwave processing to polymers and polymer composites*

Dr. Raymond Decker, University Science Partners, *utilizing microwaves in the former Soviet Union and the application of microwave processing to a number of problems*

Dr. Leonard Dauerman, New Jersey Institute of Technology, *using microwaves to treat hazardous waste*

Mr. Edward Daniels, Argonne National Laboratory, *dissociation of hydrogen sulfide, using microwave energy*

Speakers at the December 14, 1992 meeting included:

Dr. Ed Neas, The Rubbright Group, *use of microwave heating in chemistry, with a special emphasis on analytical laboratory applications*

Mr. Hal Kimrey, Oak Ridge National Laboratory, *process scaling, technology transfer, and temperature measurement techniques used at the Oak Ridge National Laboratory*

Special thanks to Arthur C. Lind of McDonnell Douglas Research Laboratories who provided valuable information on microwave processing of polymeric composites.

Government liaison representatives briefed the committee on the programs and needs of their respective agencies:

Mr. John Walkiewicz, Bureau of Mines' Reno Research Center, *the use of microwaves for drying, comminution, and fragmentation*

ACKNOWLEDGMENTS

Dr. Alan Dragoo, Department of Energy Basic Sciences Office, *microwave programs at Oak Ridge and Los Alamos National Laboratory*

Dr. Walter Zukas, Army Materials Laboratory, *microwave processing of polymers, ceramics, and thick section polymeric composites*

Dr. William Messick, Office of Naval Technology, *Materials processing needs of the Navy*

Dr. Sunil Dutta, National Aeronautics and Space Administration, Lewis Research Center, *needs in ceramics, ceramic composites and composites*

Dr. William Coblenz, Advanced Research Project Agency, Defense Science Office, *microwave processing of structural ceramics, CVD of ceramics and diamond, and curing of conducting polymer adhesives and polymeric composites*

The committee is grateful for their contribution to defining the scope of the study and their active participation in the work of the committee.

The Chair thanks the members of the committee for their efforts. Many made in-depth presentations on special topics, and all contributed to the writing and rewriting of the report. Through it all they remained congenial, hard working, and committed to a balanced and objective report. A Chair could not ask for a better group of people.

Special thanks go to Aida Neel, who made the many arrangements necessary for productive and pleasant meetings, and Tom Munns, the National Materials Advisory Board program officer; whose dedication, good humor, and steady prodding kept the report on schedule.

ABSTRACT

The use of microwaves in industrial materials processing can provide a versatile tool to process many types of materials under a wide range of conditions. Microwave processing is complex and multidisciplinary in nature and involves a wide range of electromagnetic equipment design and materials variables, many of which change significantly with temperature. A high degree of technical and other (e.g., economic) knowledge is required in determining how, when, and where to use microwaves most effectively, and when not to use them.

The committee conducted an assessment of the potential of microwave technology for industrial applications. This assessment included a review of microwave technology, equipment, processing methods, and applications. Barriers to industrial applications and gaps in understanding of microwave processing technology were identified, as were promising applications and development opportunities that take advantage of unique performance characteristics of microwaves.

ABSTRACT

PREFACE

The microwave processing of materials is a relatively new technology that provides new approaches to improve the physical properties of materials; provides alternatives for processing materials that are hard to process; reduces the environmental impact of materials processing; provides economic advantages through the saving of energy, space, and time; and provides an opportunity to produce new materials and microstructures that cannot be achieved by other methods.

Microwave processing is an unusual technology. It is widely used (more than 60 million home units are used to cook food) in an environment in which the user understands little of the technology. Yet, the difficulty in applying the technique in industrial processing has often lead to frustration of technically competent materials processors.

Some of the mystery of microwave processing is associated with this dichotomy. If it is so easy to "nuke" a meal, it must certainly be as easy to sinter a ceramic or cross-link a polymer. The apparent ease in using microwaves in food processing is a tribute to the equipment manufacturers for their success in making a "user friendly" oven, but these ovens also have the advantage of having a molecule widely distributed in the food, the water molecule, that easily "couples" to microwaves. In materials processing coupling of microwave energy must be to atoms or atomic groups other than water, at much greater technical complexity.

The purpose of this report is to:

- Introduce the reader to the use of microwaves for processing materials. The basic interactions will be described, along with the basic equipment required to process materials. Examples of successful applications will be presented, as will an evaluation of the conditions or parameters needed for the successful application of microwaves to the processing of materials.
- Provide an assessment of the state-of-the-art of microwave processing as an industrial technology.
- Identify gaps, limitations, or weaknesses in the understanding of the use of microwaves in materials processing, and suggest research and development to address these issues.

The committee approached its responsibility to evaluate the potential of microwave processing of materials in a critical and objective way. To some the report may seem overly cautious and to others it may seem overly critical. It is very clear that the microwave processing of materials has had some major advantages and major successes. But it is equally clear that potential users should take the time to become knowledgeable about microwaves and their interaction with materials before embarking on a program of using microwaves to process materials. It is the committee's hope that this report will promote the successful application of microwave processing to real-world problems.

DALE F. STEIN
CHAIR

CONTENTS

	EXECUTIVE SUMMARY	1
1	INTRODUCTION	5
	Perspective,	6
	Materials Interaction,	7
	Other Considerations,	8
2	MICROWAVE FUNDAMENTALS	9
	Microwave Generators,	10
	Candidate Generators,	12
	Wave Propagation,	18
	Waveguide Modes,	22
	Interactions Between Microwaves and Materials,	27
3	MICROWAVE SYSTEM INTEGRATION	39
	Microwave Applicators,	39
	Microwave Safety Standards,	46
	Temperature Measurements,	49
	Computer Modeling and Computer Simulation,	59
4	APPLICATION CRITERIA	67
	Unique Performance Characteristics,	67
	Economics of Microwave Processing,	71
5	MICROWAVE APPLICATIONS	79
	Introduction,	79
	Ceramics/Ceramic Matrix Composites,	80
	Polymers and Polymer Matrix Composites,	98
	Microwave Plasma Processing of Materials,	105
	Minerals Processing,	107
	Microwave Chemistry,	108
	Waste Processing and Recycling,	112
	Summary,	115

6	CONCLUSIONS AND RECOMMENDATIONS	117
	Applications Development,	118
	Process Modeling and Simulation,	118
	System Design and Integration,	119
	Nonthermal Microwave Effects,	120
	REFERENCES	121
	APPENDIX	149

EXECUTIVE SUMMARY

Microwave processing of materials is a technology that can provide the material processor with a new, powerful, and significantly different tool to process materials that may not be amenable to conventional means of processing or to improve the performance characteristics of existing materials. However, due to the complexity of microwave interactions with materials, simply placing a sample in a microwave oven and expecting it to heat efficiently will seldom lead to success.

Microwaves are electromagnetic waves in the frequency band from 300 MHz (3×10^8 cycles/second) to 300 GHz (3×10^{11} cycles/second). Industrial microwave processing is usually accomplished at the frequencies set aside for industrial use, 915 MHz, 2.45 GHz, 5.8 GHz, and 24.124 GHz.

First controlled and used during the second world war in radar systems, the usefulness of microwaves in the heating of materials was first recognized in 1946. Raytheon introduced the first microwave oven to the marketplace in 1952. During the past two decades, the microwave oven has become a ubiquitous technology, present in more than 60 million homes. Despite this long history and widespread use, there still remains a great deal that is not fully understood about microwaves and their use.

The Department of Defense and the National Aeronautics and Space Administration requested that the National Materials Advisory Board of the National Research Council conduct a study to (1) assess the current status of microwave processing technology; (2) identify applications of microwave technology where resulting properties are unique or enhanced relative to conventional processing or where significant cost, energy, or space savings can be realized; and (3) recommend future activities in microwave processing. The Committee on Microwave Processing of Materials: An Emerging Industrial Technology was established to conduct this study.

A large investment has been made over many years in the development of microwave processing systems for a wide range of product applications. In general, microwave processing systems consist of a microwave source, an applicator to deliver the power to the sample, and systems to control the heating. Microwave generators are generally vacuum tubes, but solid state devices are sometimes used. The magnetron is the most common microwave source in materials processing applications. Microwave energy is applied to samples via microwave applicators. The most common applicators are multimode (e.g., home ovens), where numerous modes are excited simultaneously, and single-mode, where one resonant mode is excited.

Control of temperature in microwave heating processes is generally accomplished through variation of input power or through pulsed sources.

Microwaves possess several characteristics that are not available in conventional processing of materials, including:

- penetrating radiation;
- controllable electric field distributions;
- rapid heating;
- selective heating of materials through differential absorption; and
- self-limiting reactions.

These characteristics, either singly or in combination, present opportunities and benefits that are not available from conventional heating or processing methods and provide alternatives for the processing of a wide variety of materials, including rubber, polymers, ceramics, composites, minerals, soils, wastes, chemicals, and powders. The characteristics of microwaves also introduce new problems and challenges, making some materials very difficult to process. First, bulk materials with significant ionic or metallic conductivity cannot be effectively processed due to inadequate penetration of the microwave energy. Second, insulators with low dielectric loss factors are difficult to heat from room temperature due to their minimal absorption of the incident energy. Finally, materials with permittivity or loss factors that change rapidly with temperature during processing can be susceptible to uneven heating and thermal runaway. While the use of insulation or hybrid heating can improve the situation, stable microwave heating of these types of materials is problematic.

The committee found that efforts in microwave process development that succeeded commercially did so because there was a compelling advantage for the use of microwave energy. Failure almost always resulted from simple, general causes e.g., trying to process materials that were not conducive to microwave absorption or trying to use equipment that was not optimized for the particular material and application.

The most likely candidates for future production-scale applications will take full advantage of the unique characteristics of microwaves. For example, chemical vapor infiltration of ceramics and solution chemical reactions are enhanced by reverse thermal gradients that can be established using microwaves. Polymer, ceramic, and composite joining processes and catalytic processes are enabled by selective microwave heating. Powder synthesis of nanoparticles can take full advantage of rapid microwave heating to produce unique formulations and small particle sizes. Thermoplastic composite lamination and composite pultrusion processes are enhanced by rapid and bulk heating and by the ability to tailor the material's dielectric properties to microwave processes. The potential for portability and remote processing also make microwave processing attractive for waste remediation.

Due to the high cost of microwave generators and the relatively poor efficiency of electric power for heating applications, factors other than energy generally account for savings realized from microwave processing. Such factors include process time savings, increased process yield, and environmental compatibility.

KEY FINDINGS AND CONCLUSIONS

The future of microwave processing of materials appears to be strongest in specialty applications, and it will probably be of limited usefulness as a general method of producing process heat. Within the specialized areas, microwave processing has distinct advantages over conventional processing means. Microwave processing will not be applicable to all materials and in fact may be readily applicable only to certain types of materials.

Failure to realize expected benefits from microwave processing is a result of inadequate interaction among researchers, materials engineers, process designers, and microwave engineers. In most cases, the basic equipment (e.g., generators, applicators, power supplies) for microwave processing applications is commercially available. However, the methodology for system integration, including system design, special applicator design, rapid equipment prototyping, and process control, is inadequate. It must be recognized that samples cannot be heated efficiently and uniformly if simply placed in a microwave oven without consideration of specific microwave/materials interactions.

The development of hybrid heating systems that optimally combine microwave sources with conventional sources to balance process variables such as required power, process flow time, tooling requirements, etc., represents a very promising, largely untapped area in process development. Hybrid heating may be provided actively, using a separate conventional heat source, or passively, using higher dielectric loss susceptors, insulation, or coatings that more readily absorb the incident power. Development of hybrid heating systems may be required for full realization of the benefits of microwave technology.

Most of the current research has focused on laboratory-scale, exploratory efforts. In order to realize the potential benefits of microwave and hybrid processes, work is needed to scale-up process and system designs to large-batch or continuous processes. Process scaling includes model simulation, system design and integration, and an understanding of the costs and benefits involved in moving to production scale.

An important element of microwave process development and system design is the capability to model electromagnetic interactions. An understanding of the variation of dielectric properties with temperature and processing state is crucial for simulations and process modeling. Computer modeling can be used to optimize generator or applicator system design, establish achievable processing windows, and conduct realistic process simulations for given dielectric properties, sample size, and desired processing conditions.

Although there is evidence of enhancements of processes due to the effects of microwaves alone (e.g., enhanced ceramic sintering, grain growth, and diffusion rates and faster apparent kinetics in polymers and synthetic chemistry), the evidence is equivocal due to questionable temperature measurement techniques, uncertain process characterization methods, and conflicting evidence.

RECOMMENDATIONS

- For particular materials, define the conditions under which microwaves provide uniform, stable processing. These may be developed through appropriate numerical modeling techniques and should be presented as processing charts that contain information on material properties, processing conditions, and specimen size and geometry. This modeling requires characterization of the thermal and physical properties of materials, including thermal conductivity and diffusivity, thermal expansion, and the temperature-dependent dielectric properties. Hybrid heating schemes, in which traditional heating is augmented with microwave heat, should be considered.
- Emphasize research work that facilitates the transition of developmental processes to production scale. This may include materials property characterization, process simulation, control schemes, equipment prototyping, and pilot-scale production.
- Establish multidisciplinary teams, consisting of materials and process engineers, microwave engineers, equipment designers, and manufacturing specialists, to properly develop microwave processes and procedures.
- Provide training in fundamentals of microwave processing technology, including microwave interactions with materials.
- Define general specifications for applicator design, and characterize the resulting electromagnetic field to enable users to successfully apply microwaves to materials processing.
- Compile existing material-property information on dielectric, magnetic, and thermal properties (including dependence on frequency and temperature) in the range useful in the processing of materials.
- Provide more-complete and more-consistent measurements of basic dielectric properties of materials to be processed using microwaves, and develop calibration standards for comparing the various techniques for dielectric properties measurements.
- Develop empirically simplified models and "microwave heating diagrams" based on measurements and on the extensive data collected from results of numerical simulation to make numerical techniques more accessible to processors.
- Establish standards for measurement of temperature to ensure reproducibility. In addition, the techniques and procedures used to measure temperature should be reported in detail, so an evaluation of accuracy can be made. The level of uncertainty in temperature measurements should also be reported. Perform experiments using several techniques for measuring temperature to determine the relative accuracy and reproducibility of the various techniques against a known standard (melting point, phase transition temperature, etc., of well characterized materials).
- Develop practical methods to monitor or determine internal temperature and thermal profiles (thermal gradients) within a material during the process cycle.
- Conduct detailed and controlled experiments to determine if the "microwave effect" reported for materials is valid. Care should be taken to use a microwave source with predictable and reproducible fields and to have an internal temperature calibration to avoid temperature measurement uncertainties.

1
INTRODUCTION

The purpose of this report is to (1) assess the current status of microwave processing technology; (2) identify applications of microwave technology where resulting properties are unique or enhanced relative to conventional processing or where significant cost, energy, or space savings can be realized; and (3) recommend future Department of Defense and National Aeronautics and Space Administration research and development activities in microwave processing.

Microwave processing of materials is a developing technology that has proven useful in a number of applications. However, microwave processing has not always been as successful as proponents of the technology had hoped. Due to the complexity of microwave interactions with materials, the successful application of microwave processing places a heavier demand on the user to understand the technique than does conventional heating. The blind application of microwave processing will likely lead to disappointment; however, wise application may have greater advantages than had been anticipated. Materials processors are becoming more sophisticated at tailoring the material to the manufacturing process in order to make full use of the capabilities of microwaves.

Much of the body of research in microwave processing of materials is exploratory in nature, often applying to particular materials, sample sizes and geometries, equipment, and processing methods. While providing valuable empirical information, these studies have not advanced the general understanding of microwave processing as much as studies more grounded in the fundamental interactions between microwaves and materials that could more readily lead to scalable, repeatable production processes. In this report, the committee seeks to develop an understanding of microwave processing, starting with fundamental interactions, that includes process development, equipment selection and design, materials evaluation, and applications development.

This report has two primary goals. The first is to provide the information needed for a basic understanding of microwave processing technology and of the strengths and limitations of microwave processing in order to assist materials processors in making wise decisions in using microwaves. Examples of successful applications will be presented to help create an understanding of the conditions necessary for success. The second goal is to identify research and development that will be crucial to the enhancement of microwave processing of materials.

Due to the broad range of industries, materials, and processes involved in the application of microwaves, the committee limited the scope of the study to materials processing, especially with regard to industrial materials and advanced materials processes of interest to Department of Defense and the National Aeronautics and Space Administration. Due to the availability of commercial equipment for more mature applications and review articles of previous research, more recent work is emphasized. Developments in these more mature application technologies such as microwave processing of rubber and plasma processing, are reviewed only briefly, emphasizing aspects important to developing commercial applications in other areas. In the case of plasma processing, applications or developments where the use of microwave-frequency plasma had specific benefit were emphasized. The application of microwave processing in wood products, biomedical, pharmaceutical, and food processing industries is not included in the committee's assessment.

PERSPECTIVE

High-frequency heating really started when engineers working onshort-wave transmitters contracted artificial fevers. The great virtuesof this kind of heat are as follows: The heat is generated directlyin the object itself; no transfer of heat is involved. Associated apparatus need not be heated. The surfaces of the material need not be affected. The people who work with the equipment have cooler working conditions. No gases are involved and thus the likelihood of corroded surfaces is eliminated. The material can be heated from the inside-out. Finally, objects of unusual size or shape can be heated.
Scientific American, 1943

It may be useful to provide the reader with some perspective concerning microwave processing in order to facilitate understanding of the more detailed and complete discussion in subsequent chapters. Microwaves were first controlled and used during the second world war as a critical component of radar systems. Although, as described above, the virtues of radio-frequency heating were forecasted earlier, the usefulness of microwaves in the heating of materials was first discovered in 1946, and in 1952 Raytheon introduced the first microwave oven to the marketplace. During the past two decades, the microwave oven has become a ubiquitous technology, with more than 60 million homes having one. Despite this long history and widespread use, there still remains a great deal that is not understood about microwaves.

The principal problems have to do with a lack of understanding, especially by the users, of the basic interactions that occur between materials and microwaves, of the design of equipment to meet the needs of a specific application, and of the inherent limitations (including cost) of microwaves as a processing methodology. Specialists in microwave technology are hindered by an incomplete understanding of some of the basic interactions that occur between materials and microwaves, and they have an incomplete data base to test their theories and models and to provide guidance in designing proper systems for practical use. There is also the barrier between the people who understand electromagnetism, wave theory, and microwaves and the materials specialists who are, in general, limited in their understanding of electromagnetic wave theory and therefore inhibited in utilizing the technology. This report is intended to alleviate these inhibitions by explaining the fundamentals of microwave interaction with materials and by describing and explaining the systems that are used to apply the microwaves to materials, both in a way that will promote an understanding of the basics for the materials processor.

MATERIALS INTERACTION

When an electric field interacts with a material, various responses may take place. In a conductor, electrons move freely in the material in response to the electric field, and an electric current results. Unless the material is a superconductor, the flow of electrons will heat the material through resistive heating. However, microwaves will be largely reflected from metallic conductors, and therefore such conductors are not effectively heated by microwaves. In insulators, electrons do not flow freely, but electronic reorientation or distortions of induced or permanent dipoles can give rise to heating. The common experience of using microwaves to heat food is based primarily on the dipole behavior of the water molecule in the food and the dipole's interaction with microwaves. Because microwaves generate rapidly changing electric fields, these dipoles rapidly change their orientations in response to the changing fields. If the field change is occurring near the natural frequency at which reorientation can occur, a maximum in energy consumed is realized, and optimum heating occurs. In the terminology of microwave processing, when this happens it is said the material is well "coupled."

The material properties of greatest importance in microwave processing of a dielectric are the permittivity (often called the dielectric constant), ϵ, and the loss tangent, $\tan\delta$. A more thorough discussion of materials properties and interactions with microwaves is included in Chapter 2. The complex permittivity is a measure of the ability of a dielectric to absorb and to store electrical potential energy, with the real permittivity, ϵ', characterizing the penetration of microwaves into the material and the loss factor, ϵ'', indicating the material's ability to store the energy. The most important property in microwave processing is $\tan\delta$, indicative of the ability of the material to convert absorbed energy into heat. For optimum coupling, a balanced combination of moderate ϵ', to permit adequate penetration, and high loss (maximum ϵ'' and $\tan\delta$) is required.

The trick in microwave processing is to find a material that is polarizable and whose dipoles can reorient rapidly in response to changing electric field strength. Fortunately, many materials satisfy these requirements and are therefore candidate materials for microwave processing. However, if these materials possess poor thermal conductivity, heat does not rapidly dissipate into the surrounding regions of the material when a region in the solid becomes hot. This difficulty is compounded, because the dielectric loss frequently increases dramatically as the temperature increases. Thus, the hot region becomes even hotter, sometimes resulting in local melting. These "hot spots" are a major difficulty and have led to the use of hybrid systems, combining microwave heating with other heat sources to reduce uneven heating.

INTRODUCTION

OTHER CONSIDERATIONS

There is another consideration that is often overlooked in microwave processing of materials. Microwaves are generated by devices requiring electrical energy. Electrical energy is generated primarily from fossil fuels. The conversion of the energy in the fuel to electrical energy is less than 40 percent efficient. In addition, microwave generators (magnetrons, etc.) are not generally better than 85 percent efficient in converting electric power to microwaves, and the microwaves are not perfectly coupled to the material (90 percent coupling would be very good), so the total energy generated is probably less than 30 percent of the energy content of the fossil fuel used in generating the electricity. This means there are real limitations to the economics of bulk heating. Direct heating with fossil fuels makes much more efficient use of energy, and microwaves can only be economically competitive when electric heating is mandated or the selective heating capability of microwaves, or some other factor, more than compensates for the inefficiency of electric heating. An example, discussed later in this report, is the removal of volatiles from soil, where it is not necessary to heat the soil as is required when heating by conventional means.

The successful use of microwaves requires the processor to have a good understanding of the strengths and limitations of microwaves. Among the strengths are penetrating radiation, controllable electric field distributions, rapid heating, selective heating, and self-limiting reactions. But, simply putting a material into the microwave oven and "zapping" it in the hopes of solving a problem is risky. The materials processor must understand and match the special capabilities of microwave processing to the material and the properties required in order to design an appropriate process. In some cases, incomplete understanding exists, requiring research to improve the knowledge base for using microwaves to process materials.

The information contained in this report is intended to ease the work of those interested in using microwaves to solve a problem or improve a current process. Examples of successful applications are given to illustrate the characteristics of a material and process that are amenable to microwave processing. An equipment section describes the alternatives available and the "setup" required to apply the microwaves to the material. Economic considerations are described, and where possible, costs are provided as an aid in determining the economic consequences of using microwaves. A theory section is provided to help both the materials processor and the equipment designer understand the fundamental limitations and advantages of microwaves in the processing of materials. In addition, the limitations in present understanding are delineated as a caution to users and as a guide for future research activities.

2
Microwave Fundamentals

The initial surge in microwave technology development was driven by the military needs of the second world war. The tremendous effort that went into development of radar during World War II generated a great body of knowledge on the properties of microwaves and related technologies. Much of this information was documented in a series of volumes edited by Massachusetts Institute of Technology Radiation Laboratory under the supervision of the National Defense Research Committee (MIT, 1948; Ragan, 1948).

A basic understanding of microwaves and their interaction with materials is required to realize the promise, as well as to understand the limitations, of microwave processing. Although there is a broad range of materials that can be processed using microwaves, there are fundamental characteristics and properties that make some materials particularly conducive to microwave processing and others difficult. While an empirical understanding of microwave processing is important in moving developmental processes into production, a more fundamental approach is required for development of optimized process cycles, equipment, and controls. For instance, repeatability of a measurement is challenging in microwave processing since the results can be affected by a myriad of factors, including moisture content, changes in dielectric properties during processing, electromagnetic interference with temperature measurements, sample size and geometry, or placement of the sample within the cavity.

The purpose of this chapter is to discuss, in general terms, how microwaves are generated, introduce the fundamental nature of microwaves, how they interact with materials, and how these interactions generate process heat. The application of these fundamental concepts for design or selection of a practical processing system is discussed in Chapter 3. The unique performance characteristics that arise from the interactions of microwaves and materials and how they may be used to develop application criteria are described in Chapter 4.

The spectrum of electromagnetic waves spans the range from a few cycles per second in the radio band to 10^{20} cycles per second for gamma rays (Figure 2-1). Microwaves occupy the part of the spectrum from 300 MHz (3×10^8 cycles/s) to 300 GHz (3×10^{11} cycles/s). Typical frequencies for materials processing are 915 MHz, 2.45 GHz, 5.8 GHz, and 24.124 GHz.

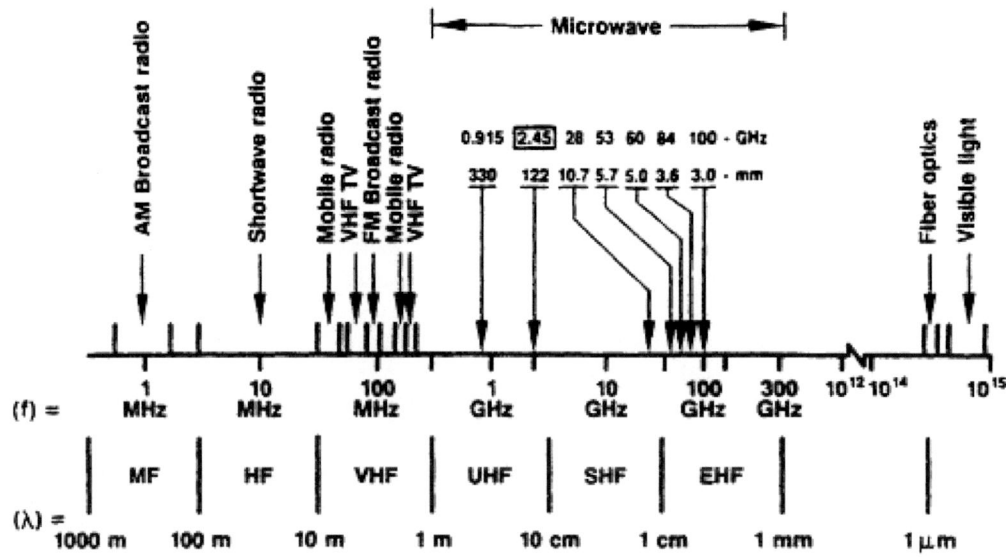

FIGURE 2-1
Electromagnetic spectrum and frequencies used in microwave processing (Sutton, 1993).

MICROWAVE GENERATORS

Major advances in microwave generation and generators occurred in the early 1940s with the invention, rapid development, and deployment of the cavity magnetron on the heels of the earlier (1938) invention of the klystron. What started as "flea powered" curiosities are now capable of generating hundreds of megawatts of power.

The genesis of microwave generator development was the introduction of DeForest's audion tube, the first electron tube amplifier, in 1907. In addition to being developed to be capable of generating megawatts of continuous-wave power (albeit limited in operating frequencies to the region below 1,000 MHz), the audion tube led to the lineage of microwave tubes. Figure 2-2 traces, in time, that beginning and the inventions and developments that followed. The dependence of one tube on the understanding and development of others is illustrated.

A historical discussion of microwave generators would be incomplete without highlighting two major developments. The first involved space charge and the transit time of electron motion within a vacuum, which represented a fundamental limitation to the operating frequency and output power of conventional gridded tubes. When the time of transit became an appreciable part of a microwave frequency cycle, performance degraded, forcing the designer to smaller and smaller sizes to achieve higher frequency. The invention of the kylstron obviated this limitation by utilizing space-charge effects *and* transit-time effects in a device whose dimensions encompass many wavelengths. A klystron today is capable of peak output power of 100 megawatts at 10 GHz.

MICROWAVE FUNDAMENTALS

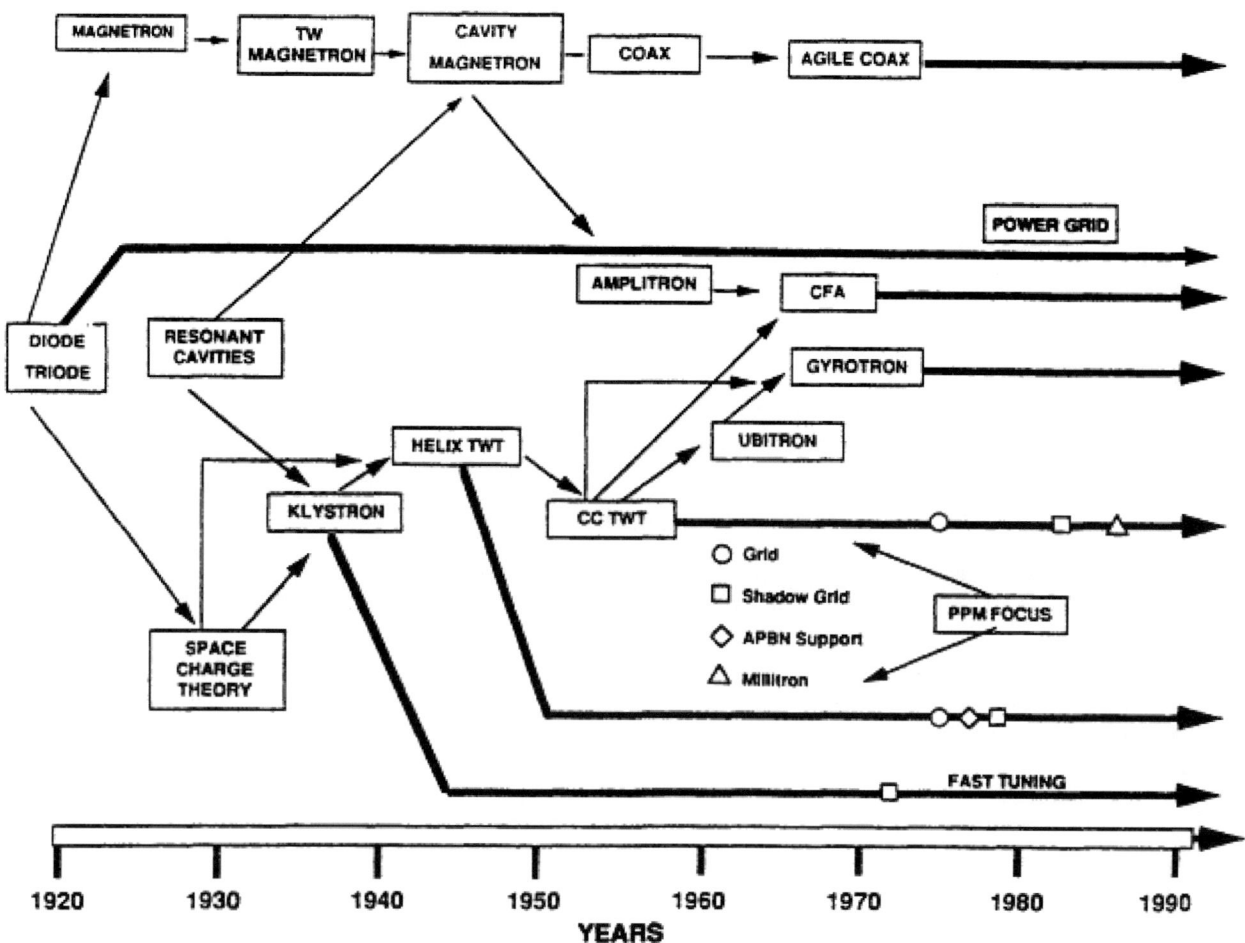

FIGURE 2-2
Microwave Tube Development.

At even higher frequency and higher power levels, limitations associated with voltage and size were encountered. The second major invention occurred in Russia with the realization that the citron relativistic mass change at high voltages could be fundamental to a new type of beam—wave interaction. This led to the development of the gyrotron, later brought to a high state of development in this country. Power/frequency goals of 1 MW continuous wave at 140 GHz are being pursued.

Large performance improvements have been achieved through the application of new materials and processes in microwave generators. For example, the application and ready availability of high thermal conductivity beryllium oxide or boron nitride has allowed signifier improvements in maximum continuous wave power output of traveling wave tubes (from approximately 3 W to 3 kW).

With roots in military radar, today's microwave tubes find applications in medical, scientific, broadcast, communication, and industrial equipment.

CANDIDATE GENERATORS

It is instructive to show device performance range on a power—frequency plot, as in Figure 2-3.

In addition to power and frequency, other performance factors are important to specific applications. Gain, linearity, noise, phase and amplitude stability, coherence, size, weight, and cost must also be considered. Currently available microwave generators include power grid tubes, klystrons, klystrodes (a combination of tetrode power grid tube and klystron), magnetrons, crossed-field amplifiers, traveling wave tubes, and gyrotrons. Those most applicable to materials processing are described.

Table 2-1 shows the most likely candidate tubes, together with a few salient characteristics, including device cost and cost per watt of power generated. The cost of the ancillary equipment such as power conditioning, control circuitry, transmission line, and applicator must be added to the numbers shown. A discussion of cost issues for microwave processing is included in Chapter 4.

Magnetron

At the customary microwave frequencies, the magnetron is the workhorse, the economic product of choice for the generation of "raw" power. These are the tubes used in conventional microwave ovens found in almost every home (with power on the order of a kilowatt in the 2—3 GHz range) and in industrial ovens with output to a megawatt.

Radars employing magnetrons number in the tens of thousands, and household ovens employing the so-called "cooker magnetron" number in the tens of millions. Large quantities often lead to lower cost, and thus for many microwave heating and processing applications the magnetron is the device of choice, with advantages in size, weight, efficiency, and cost.

The magnetron is the major player in a class of tubes termed "crossed field," so named because the basic interaction depends upon electron motion in electric and magnetic fields that are perpendicular to one another and thus "crossed." In its most familiar embodiment, shown schematically in Figure 2-4, a cylindrical electron emitter, or cathode, is surrounded by a cylindrical structure, or anode, at high potential and capable of supporting microwave fields. Magnets are arranged to supply a magnetic field parallel to the axis and hence perpendicular to the anode cathode electric field.

The interaction of electrons traveling in this crossed field and microwave fields supplied by the anode causes a net energy transfer from the applied DC voltage to the microwave field. The interaction occurs continuously as the electrons traverse the cathode anode region. The magnetron is the most efficient of the microwave tubes, with efficiencies of 90 percent having been achieved and with 70—80 percent efficiencies common. Figure 2-5 shows the conventional "cooker magnetron." Figure 2-6 is an industrial heating magnetron designed to operate at 2,450 MHz at the 8 kW level.

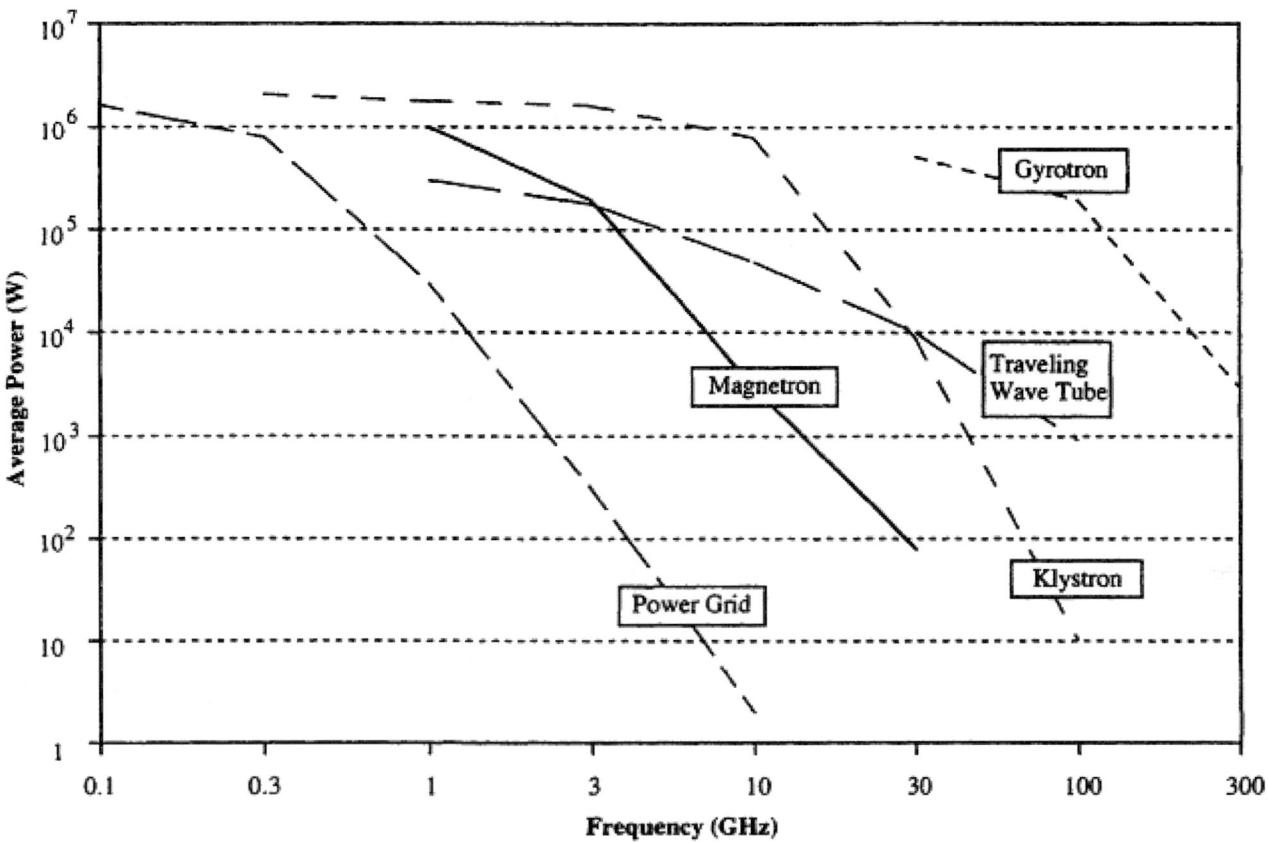

FIGURE 2-3
Power——frequency limits of microwave generators.

Power Grid Tubes

At somewhat lower frequencies, power-grid tubes (triodes, tetrodes, etc.) are selected for their lower cost. These can be the same as tubes utilized in AM and FM radio broadcasting and in television. The large user base provides the quantity factor important to economical manufacturing and hence low cost.

The power-grid robe consists of a planar element emitting electrons (cathode) near a parallel planar element receiving electrons (anode) and of an interspersed parallel grid controlling the electron flow. With proper DC voltages applied to the three elements, a microwave signal impressed on the grid results in a much larger (amplified) signal at the anode.

The performance of power-grid robes rages from low-power tubes capable of roughly a kilowatt of output at frequencies of 350 MHz to very-high-power tubes used in fusion research

TABLE 2-1 Microwave Tube Characteristics

	Power KW	Frequency GHz	Efficiency %	Cost $K	Cost $/Watt	Special Rqmts.
Magnetron						
"Cooker"	1	2.45	60-70	0.05	0.05	
Industrial	5 to 15	2.45	60-70	3.50	0.35	
Industrial	50	0.915	60-70	5.00	0.10	
Power Grid (Transmitting)						
"Low"	to 10	to 3	80s		0.1 to .4	ex
"Medium"	to 100	to 1	80s		0.10	ex
"High"	to 2000	to .15	80s		0.10	ex
Klystron						
Example	500	3	60s	350.00	0.70	s
Example	250	6	60s	200.00	0.80	s
Example	250	10	60s	375.00	1.50	s
Klystrode						
Example	30	1	70s	18.00	0.60	s, ex
Example	60	1	70s	36.00	0.60	s, ex
Example	500	1	70s	Dev.	~ 1	s, ex
Gyrotron						
Example	200	28	30s	400.00	2.00	sc
Example	200	60	30s	400.00	2.00	sc
Example	500	110	30s	500.00	1.00	sc
Example	1000	110	30s	800.00	0.80	sc
Example	1000	140	30s	Dev	~ 1	sc

Special Requirements:
sc-superconducting solenoid
s-solenoid for beam focus
ex-external circuitry, cavities, etc.

MICROWAVE FUNDAMENTALS

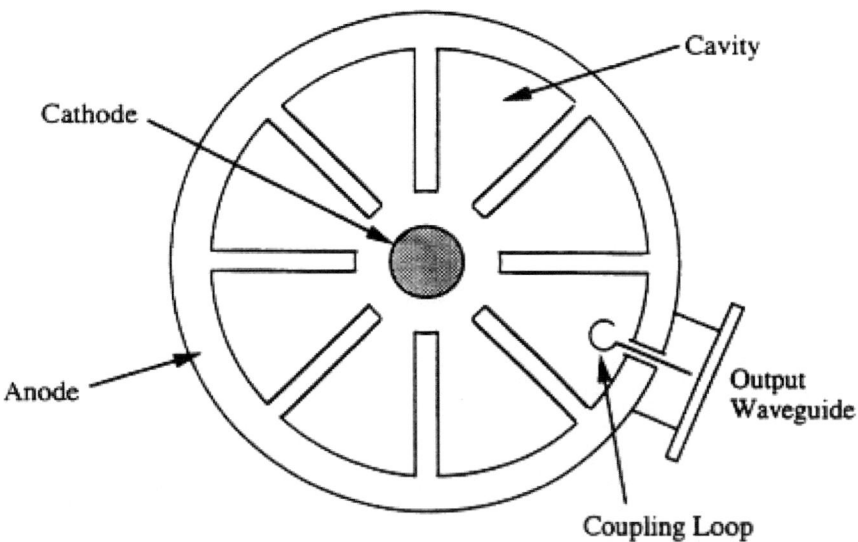

FIGURE 2-4
Schematic diagram of a magnetron shown in cross-section.

FIGURE 2-5
"Cooker" magnetron used in home microwave oven.

(plasma heating) and capable of more than 2 MW of output in the 30 MHz range. Such tubes achieve their capability through application of pyrolytic graphite grid structures, an array of which is shown in Figure 2-7.

FIGURE 2-6
Industrial heating magnetron. (Photo courtesy of English Electric Valve Company, Limited).

Specialty Tubes (Klystron, Gyrotron)

The other classes of microwave tubes will be selected for use only for those specialized requirements not fulfilled by the magnetron or power-grid tube, that is, when there is a need for very high continuous wave power (klystron) or very high continuous wave power at very high frequency (gyrotron).

Klystrons range in length from an inch or two to as long as 25 feet and operate at voltages from a few hundred to several hundred thousand volts. They power most modern radars (both civilian and military), special material processing equipment, and the linear accelerators used in science (e.g., at the Stanford Linear Accelerator Center) and medicine (e.g., for radiation cancer therapy). Efficiencies as high as 80 percent have been achieved, although 50—60 percent is more common.

FIGURE 2-7
Pyrolytic graphite grid structures for power grid-tubes.

The gyrotron is large, heavy, and expensive and employs very high voltages and high magnetic fields generated by superconducting solenoids. However, the can do what no other robe has done——generate very high power levels at millimeter wavelengths. Much of the initial gyrotron development was sponsored by the Department of Energy to find a source for plasma hinting in fusion reactors (tokamak). Gyrotrons are driving such plasmas in Japan, Europe, and the former Soviet Union, and in the United States at Princeton University and General Atomics.

Material processing is able to benefit from this long and expensive development. Equipment operating at 28 GHz, 60-kW continuous wave power is available at Oak Ridge National Laboratory as well as Tokyo, Japan. Much higher power and frequency have been demonstrated and could be made available if the technical and economic studies warrant the investment. Currently, high-power tubes are operating at 60 GHz, 110 GHz, and 140 GHz. A large experimental gyrotron has achieved 500 kW at 110 GHz.

WAVE PROPAGATION

Microwave propagation in air or in materials depends on the dielectric and the magnetic properties of the medium. the electromagnetic properties of a medium are characterized by complex permittivity (ϵ) and complex permeability (μ), where:
and

$$\epsilon = \epsilon' - i\epsilon''$$
$$\mu = \mu' - i\mu''$$

The real component of the complex permittivity, ϵ', is commonly referred to as the dielectric constant. since, as will become evident in this report, ϵ' is not constant but can vary significantly with frequency and temperature, it will be referred to simply as the permittivity (Risman, 1991). The imaginary component of complex permittivity, ϵ'', is the dielectric loss factor. similarly, the real and imaginary components of the complex permeability, μ' and μ'', are the permeability and magnetic loss factor, respectively.

Wave Equations and Wave Solutions

In wave propagation, if z is the direction of wave propagation and t is the time, the amplitude of the electric field and that of the magnetic field vary sinusoidally in both z and t. The number of complete cycles in a second is the frequency, f, and the distance that the wave travels in a complete cycle is the wavelength, λ_g. Hence, the frequency and the wavelength specify how a wave behaves in time and in distance.

It is instructive to illustrate the principles of wave phenomena by considering, in detail, a plane wave (Figure 2-8). A plane wave's wave-front is a plane normal to z with fields that are uniform in x and y. The wave has an electric field E_x in the x direction and a magnetic field, H_y in the y direction. For simplification, it is assumed that $\epsilon = \epsilon'$ and $\mu = \mu'$, since the real part of ϵ or μ is much larger than the imaginary part. Once the wave properties are calculated by this approximation, the approximation can be improved by taking the imaginary components of ϵ and μ as a perturbation.

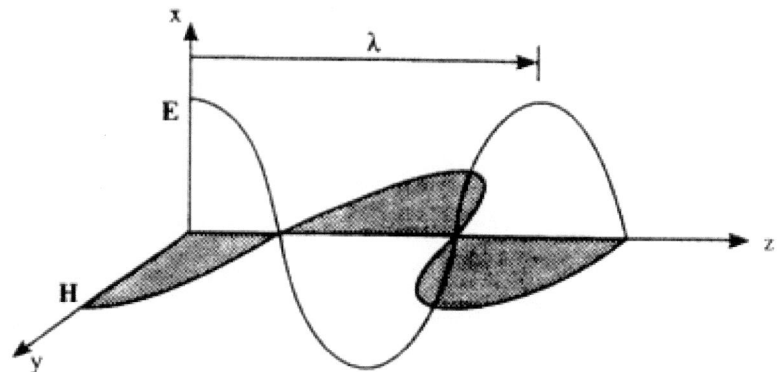

FIGURE 2-8
Propagation of a plane wave.

The wave equations for the electric and the magnetic fields derived from the Maxwell equations are

$$\frac{\partial^2 E_x}{\partial t^2} = -\omega^2 \varepsilon_0 \mu_0 \varepsilon' \mu' E_x$$

$$\frac{\partial^2 H_y}{\partial t^2} = -\omega^2 \varepsilon_0 \mu_0 \varepsilon' \mu' H_y \tag{1}$$

where ε_0 and μ_0 are, respectively, the free space permittivity (8.854×10^{-12} F/m) and free space permeability (1.256×10^{-6} H/m) in MKS units.

The solutions are

$$E_x = E_x e^{i\omega t} e^{-i2\pi z/\lambda_g}$$

$$H_y = H_y e^{i\omega t} e^{-i2\pi z/\lambda_g} \tag{2}$$

for a wave traveling forward in the + z direction and

$$E_x = E_x e^{i\omega t} e^{i2\pi z/\lambda_g}$$

$$H_y = H_y e^{i\omega t} e^{i2\pi z/\lambda_g} \qquad (3)$$

for a backward (-z) traveling wave. The wave has a time-dependence in ω, where ω is the angular frequency and is equal to $2\pi f$, where f is the frequency in cycles/second or hertz. It has a z dependence in $2\pi/\lambda_g$, which is called the wave constant.

$$\lambda_g = \frac{2\pi}{\omega(\varepsilon_0 \mu_0 \varepsilon' \mu')^{1/2}} \qquad (4)$$

If the dielectric medium is lossy, a complex ϵ must be used. Taking $\epsilon = \epsilon' - i\epsilon''$ instead of ϵ' in Equation 1, the wave solution for the forward wave is then:

$$E_x = E_x e^{-\alpha z} e^{i\omega t} e^{-i2\pi z/\lambda_g} \qquad (5)$$

where, the attenuation constant α is

$$\alpha = \tfrac{1}{2}(2\pi/\lambda_g)\varepsilon''/\varepsilon'$$

Equation 5 indicates that the amplitude of the wave decreases exponentially as it propagates i.e., wave energy is dissipated during the propagation.

For the isotropic medium considered here, one remarkable property of the wave is that it carries an equal amount of energy in the electric and magnetic fields. For the plane wave in any plane normal to z, the electric energy density is $\tfrac{1}{2}\varepsilon_0 \varepsilon' E_x^2$, and the magnetic energy density is $\tfrac{1}{2}\mu_0 \mu' H_y^2$. Since they are equal,

$$\tfrac{1}{2}\varepsilon_0 \varepsilon' E_x^2 = \tfrac{1}{2}\mu_0 \mu' H_y^2$$

Therefore,

$$\frac{E_x}{H_y} = Z = \left(\frac{\mu_0 \mu'}{\varepsilon_0 \varepsilon'}\right)^{1/2}$$

The ratio of E_x to H_y, or Z in Equation 6 is called the *wave impedance*. It characterizes the transverse electric-and magnetic-field profile of the wave and is thus also known as the *characteristic impedance*, or the *intrinsic impedance*. In air or a vacuum, $\epsilon' = \mu' = 1$ and Z is 377 ohms. Another remarkable property of microwaves is that at any plane normal to z, the electric and magnetic fields are in phase. The wave impedance is thus a real quantity and has the same value anywhere along the z axis.

The power flow in the z direction is then,

$$P_z \mathbf{z} = E_x H_y \mathbf{z} = H_x^2 Z_z \qquad (7)$$

The vector quantity, $P_z \mathbf{z}$, is the Poynting vector, which is the power flow carded by the wave. The unit vector **z** indicates that the power flow points in the z direction.

In writing the dielectric constant ϵ as a complex quantity $\epsilon' - i\epsilon''$, the real component, ϵ', affects the wavelength in Equation 4 and the impedance in Equation 6, while the imaginary component, ϵ'', represents the dielectric loss indicative of the microwave power absorbed by the medium and converted into heat.

Standing Waves

The two remarkable properties of wave propagation stated earlier are that the wave carries an equal amount of the electric and magnetic energy and that the wave impedance stays constant in the propagation direction. These are intrinsic properties of transmission lines and are true as long as the forward and the backward traveling wave are separate. However, these conditions no longer hold when both the forward and backward traveling waves exist simultaneously. Consider the case that a conducting plane is placed at $z=z_L$, causing reflection of a backward traveling wave. Adding Equation 2 and Equation 3 together and considering the fact that the electric field must be zero and the magnetic field is maximum at the conducting plane,

$$E_x = 2E_x e^{i\omega t} \sin[2\pi(z-z_L)/\lambda_g]$$

$$H_y = 2H_y e^{i\omega t} \cos[2\pi(z-z_L)/\lambda_g]$$

$$Z = \frac{E_x}{H_y} = Z_{TEM} \tan[2\pi(z-z_L)/\lambda_g] \qquad (8)$$

Both the electric and magnetic fields are in the form of a standing wave in the sense that all the energy carried by the wave forward is reflected back and that the wave is neither moving the energy forward nor backward. The wave impedance, Z, is no longer a constant and it varies as a tangent function in **z**.

Reflection and Transmission Coefficients

In the above case, a wave is reflected by placing a conducting plane at z_L. A reflection also occurs when there is a discontinuity in the dielectric or magnetic property of the medium. In this case, however, only a portion of the wave is reflected. Consider the case in which a boundary separates two media with ϵ_1, μ_1, and ϵ_2, μ_2 and wave impedances of Z_1 and Z_2, respectively.

Let an incident wave of an amplitude 1 volt/m travel forward in medium 1 normal to the boundary (Figure 2-9). Both Z_1 and Z_2 are real quantities; the reflected wave in medium 1 has an amplitude $(K-1)/(K+1)$ volt/m and the wave transmitted into the medium 2 has an amplitude $2K/(K+1)$ volt/meter, where $K=Z_1/Z_2$. The ratio $(K-1)/(K+1)$ is called the *reflection coefficient*, and the ratio $2K/(K+1)$ is called *the transmission coefficient*. At the boundary, the sum of the tangential electric (magnetic) field of the incident wave and that of the reflected wave are equal to the tangential electric (magnetic) field of the transmitted wave.

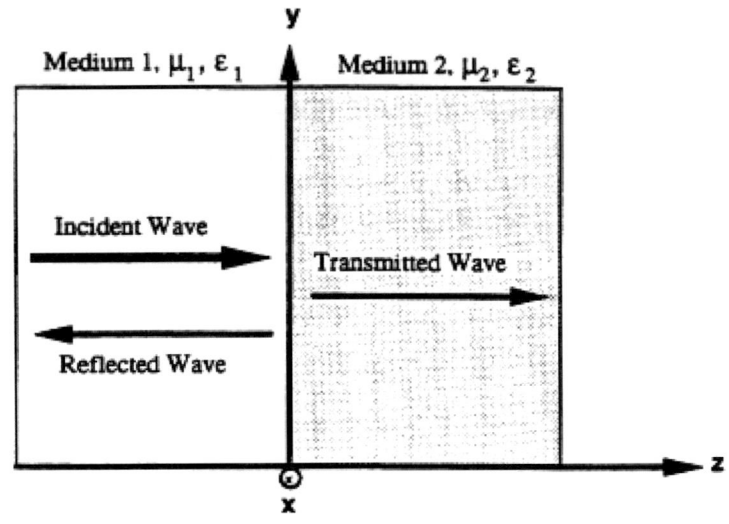

FIGURE 2-9
Schematic representation of reflection and transmission of a normal incidence electromagnetic wave.

WAVEGUIDE MODES

Microwaves can be divided into three types (Figure 2-10). For the transverse electromagnetic (TEM) wave, all fields are transverse. It is an approximation of the radiation wave in free space. It is also the wave that propagates between two parallel wires, two parallel plates, or in a coaxial line. The parallel wires, or more precisely, twisted pairs, and the coaxial lines are used in the telephone industry. The transverse electric (TE or H) wave and the transverse magnetic (TM or E) wave are those in waveguides, which are typically hollow conducting pipes having either a rectangular or a circular cross section. In the TE wave, the z component of the electric field is missing, axed in the TM wave, the z component of the magnetic field is missing. Complicating the matter further, each TE. and TM wave in a waveguide can have different field configurations. Each field configuration is called a mode and is identified by the indexes m an n. In mathematics, those indexes are the eigenvalues of the wave solution.

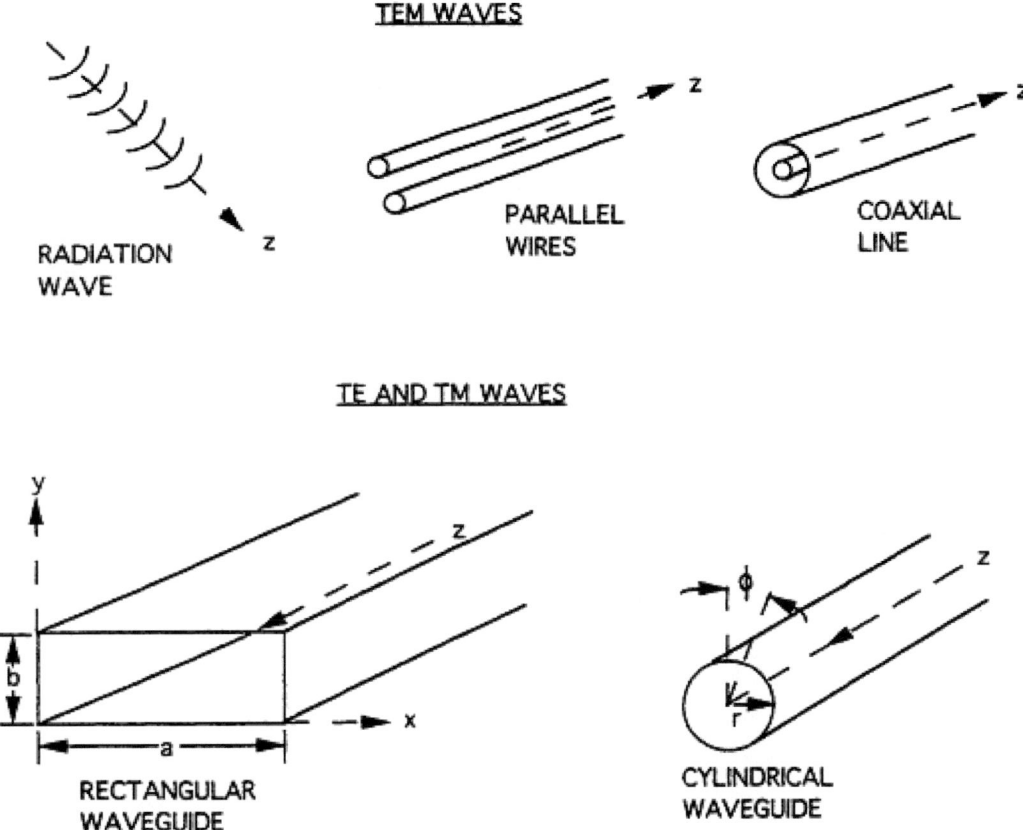

FIGURE 2-10
Waveguides - TEM, TE, and TM waves.

Field distributions for various modes of propagation in rectangular and cylindrical waveguides are available in standard text books (Ramo and Whinnery, 1944; Iskander, 1992). TE_{mn} and TM_{mn} modes are considered in rectangular waveguides and TE_{nl} and TM_{nl} modes are considered in cylindrical waveguides, where the indices m, n, and 1 are the order of the modes. Mathematically, they can be any integer, 0, 1, 2, etc. However, depending upon the size of the waveguide, physical reality allows only the lower values of m, n, and 1 and thus limits the number of the modes that can exist in a waveguide. For the rectangular waveguides, the TM wave (or E wave) has the field components E_x, E_y, E_z, H_x, and H_y, and the TE wave (or H wave), the field components E_x, E_y, H_x, H_y, and H_z. For the cylindrical waveguides, the TM wave has the field components E_r, E E_z, H_r, and H , and the TE wave, E_r, E , H_r, H , and H_z. The rectangular waveguide has a height "b" in **y** and a width "a" in **x**, and the cylindrical waveguide has a radius, "a." Wave propagation is always in the z direction. Figures 2-11 and 2-12 show the field distributions of some lower-order waveguide modes. In the figures, electric fields are represented by solid lines and magnetic fields by dashed lines. Here again, the wave carries an equal amount of the energy in electric and magnetic fields. Since not all the fields are transverse, the energy carried by the **z** component of the field must be taken into the calculation. To calculate the energy, energy density is integrated over the cross-section of the waveguide. At any plane normal to z, the transverse electric field is always normal to, and in phase with, the transverse magnetic field. The wave impedance, which depends upon the transverse fields only, is again constant anywhere along the waveguide. The two transmission-line properties described earlier for the plane (TEM) wave, then, apply equally well to the TE and TM waves.

Wavelength and the Wave Impedance

The field distributions in Figures 2-11 and 2-12 are quite complex. In general, at a conducting surface, electric field lines are normal to the surface, and magnetic field lines are parallel to it. Away from the surface, all field lines follow continuity. Before carrying out calculations of the field distributions, the wavelength and the wave impedance of a waveguide mode will be considered. A rectangular waveguide will be used as an illustration.

For simplicity it is assumed that $\epsilon = \epsilon'$ and $\mu = \mu'$. The solutions obtained in those conditions can be extended to the general case by the perturbation method when both ϵ and μ are complex. Both the wavelength and the wave impedance depend upon the dimensions "a" and "b" of the cross-section as well as ϵ' and μ' of the medium. Fortunately, the calculation can be simplified by defining a parameter k_1, which depends upon only ϵ' and μ' of the medium, and another parameter K_c, which depends upon only the dimensions "a" and "b." Thus,

$$k_1 = \omega(\epsilon'\epsilon_o\mu'\mu_o)^{1/2} \qquad (9)$$

and

$$k_c = [(m\pi x/a)^2 + (n\pi y/b)^2]^{1/2} \qquad (10)$$

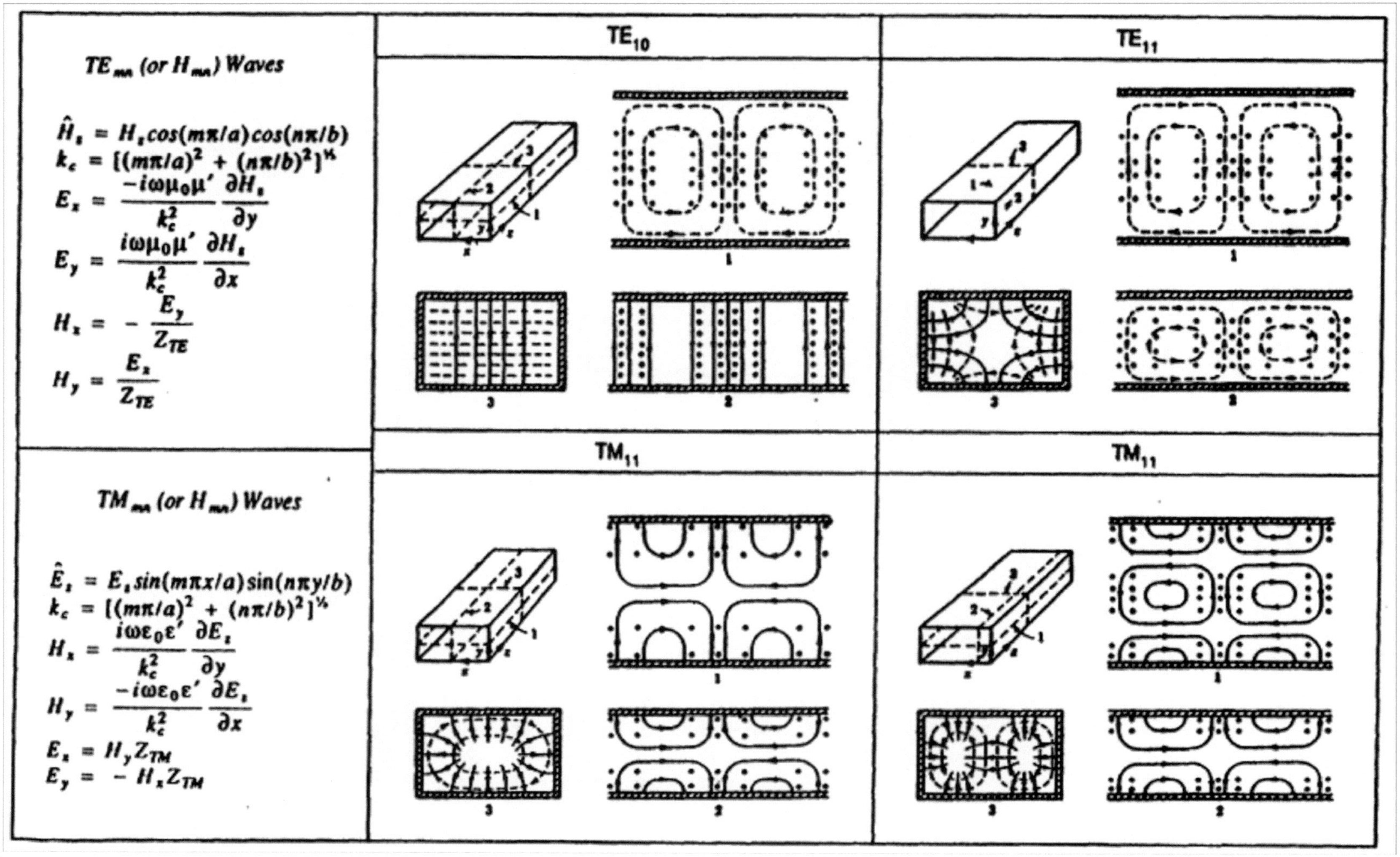

MICROWAVE FUNDAMENTALS

TE$_{mn}$ (or H$_{mn}$) Waves

$\hat{H}_z = H_z \cos(m\pi x/a)\cos(n\pi y/b)$
$k_c = [(m\pi/a)^2 + (n\pi/b)^2]^{1/2}$
$E_x = \dfrac{-i\omega\mu_0\mu'}{k_c^2} \dfrac{\partial H_z}{\partial y}$
$E_y = \dfrac{i\omega\mu_0\mu'}{k_c^2} \dfrac{\partial H_z}{\partial x}$
$H_x = -\dfrac{E_y}{Z_{TE}}$
$H_y = \dfrac{E_x}{Z_{TE}}$

TM$_{mn}$ (or H$_{mn}$) Waves

$\hat{E}_z = E_z \sin(m\pi x/a)\sin(n\pi y/b)$
$k_c = [(m\pi/a)^2 + (n\pi/b)^2]^{1/2}$
$H_x = \dfrac{i\omega\varepsilon_0\varepsilon'}{k_c^2} \dfrac{\partial E_z}{\partial y}$
$H_y = \dfrac{-i\omega\varepsilon_0\varepsilon'}{k_c^2} \dfrac{\partial E_z}{\partial x}$
$E_x = H_y Z_{TM}$
$E_y = -H_x Z_{TM}$

FIGURE 2-11
Field distributions and key expressions of calculation for modes in rectangular waveguides.

It follows that the wavelength, λ_g, and the wave impedances, Z_{TE} and Z_{TM}, for the TE and TM waves, respectively, can be expressed in terms of k_1 and K_c, such as,

$$\lambda_g = \frac{2\pi}{(k_1^2 - k_c^2)^{1/2}} \qquad (11)$$

$$Z_{TM} = \frac{(2\pi/\lambda_g)}{\omega\varepsilon_0\varepsilon'} \qquad (12)$$

$$Z_{TE} = \frac{\omega\mu_0\mu'}{(2\pi/\lambda_g)} \qquad (13)$$

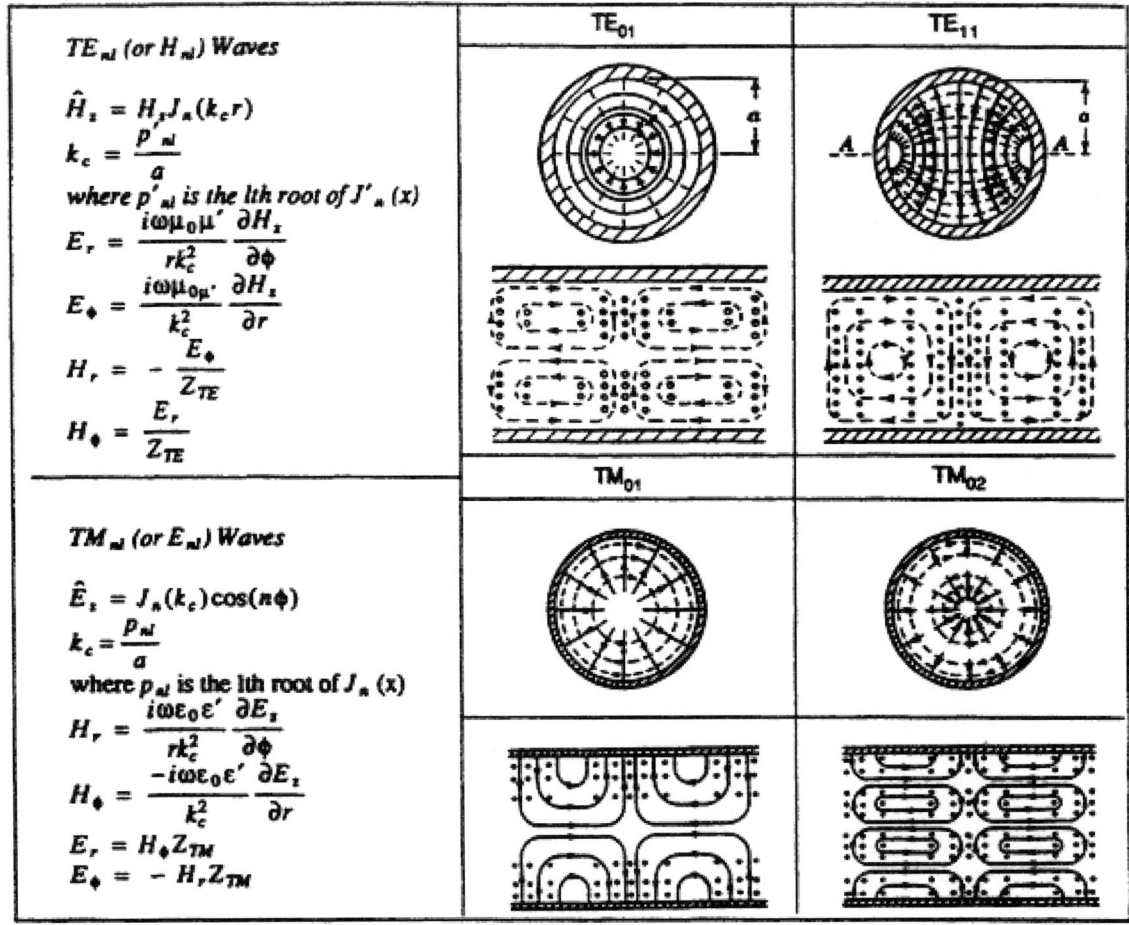

FIGURE 2-12
Field distributions and key expressions of calculation for modes in cylindrical waveguides.

The equations 9, 11, 12, and 13 above are true for all waveguides. Consider first the calculation of the wavelength in Equation 11. The wavelength must be real and positive by definition. That is true only when k_c is smaller than k_1. The waveguide mode indices, m and n, can be any integer including zero, but both cannot be zero. Since k_c involves m and n, only a limited number of combinations of the values of m and n can keep k_c smaller than k_1. The number of the modes that can propagate in a waveguide is therefore limited. For small values of "a" and "b" such that the condition $k_c < k_1$ cannot be met, no waveguide mode can exist, and the waveguide is at *cutoff frequency*. For some moderate values of "a" and "b," only one mode satisfies the condition $k_c < k_1$, and the waveguide is called a *single-mode waveguide*. For very large values of "a" and "b," thousands of the modes satisfy the condition, and the waveguide is a *multimode waveguide*.

Field Distributions

The expressions listed in the left columns of figures 2-11 and 2-12 are used to calculate field distributions. Also plotted in those figures are the field distributions of some lower-order modes in rectangular and cylindrical waveguides. Consider specifically a TM wave in a rectangular waveguide (Figure 2-11). The first expression in the "TM wave" column of Figure 2-11 is a functional form for the **z** component of the field, $\mathbf{E}_z$. The second expression is the calculation of k_c discussed earlier in Equation 10. Knowing k_c, the next two expressions in the column calculate $\mathbf{H}_x$ and $\mathbf{H}_y$. The wavelength is calculated from Equation 11, and finally, $\mathbf{E}_x$ and $\mathbf{E}_y$ are calculated from the wave impedance, Z_{TM}, given in Equation 12. The field distribution in the waveguide is thus completely determined.

INTERACTIONS BETWEEN MICROWAVES AND MATERIALS

When an electric field interacts with a material, various responses may take place. This section discusses the interactions of a variety of materials with microwave fields. Coupling mechanisms, critical electrical properties (and their variability with temperature and frequency), and the resulting conversion of incident electromagnetic energy into process heat are discussed in some detail. Also, heat transfer and related problems of uneven heating and thermal runaway are covered.

In conductors, electrons move freely in the material in response to the electric field and an electric current results. Unless it is a superconductor, the flow of electrons will heat the material through resistive heating. However, microwaves will be largely reflected from metallic conductors and therefore they are not effectively heated by microwaves.

In insulators, electrons do not flow freely, but electronic reorientation or distortions of induced or permanent dipoles can give rise to heating. The complex permittivity is a measure of the ability of a dielectric to absorb and to store electrical potential energy, with the real permittivity, ϵ', characterizing the penetration of microwaves into the material, and the loss factor, ϵ'', indicating the material's ability to store the energy. The loss tangent, $\tan\delta$, is indicative of the ability of the material to convert absorbed energy into heat. For optimum coupling, a balanced combination of moderate to permit adequate penetration and high loss (maximum ϵ'' and tanδ) are required.

Materials that are amenable to microwave heating are polarizable and have dipoles that reorient rapidly in response to changing electric field strength. However, if these materials possess low thermal conductivity and dielectric loss that increases dramatically as the temperature increases "hot spots" and thermal runaway may be experienced.

Materials considered in this report are diverse, including metal-like materials, ceramics, polymers, glass, rubber, and chemicals. The principal effect of microwave interaction is the heating of materials. The committee thus considers the conversion of microwave energy into heat, a process that involves interaction between microwave fields and the conductivity or dielectric properties of the material. Interactions between microwaves and materials can be represented by three processes: space charges due to electronic conduction, ionic polarization associated with far-infrared vibrations, and rotation of electric dipoles (Newnham et al., 1991).

The processes described above may be illustrated in combination by considering a material that has an electronic conduction σ_e, an ionic conduction σ_i, and a complex permittivity, $\epsilon' - i\epsilon''$. In the presence of an electric field, E, in the material, a current must flow.

$$E = e^{i\omega t} \; (V/m)$$

According to the Maxwell equations, the current density **j** is

$$j = [\sigma_e + \sigma_i + i\omega\epsilon_o(\epsilon' - i\epsilon'')]E$$

$$= i\omega\epsilon_o\epsilon' E + \omega\epsilon_o\epsilon' \tan\delta E \qquad (14)$$

where,

$$\tan\delta = \frac{(\sigma_e + \sigma_i)/\omega\epsilon_o + \epsilon''}{\epsilon'} \qquad (15)$$

The phase angle, δ, relates to the time lag involved in polarizing the material. The quantity tanδ is the *loss tangent, the* most important parameter in microwave processing. Returning to Equation 14, the first term at the fight of the equation is the component of the current 90 degrees out of phase with the electric field. It is the displacement current that stores electric energy in the material. The average electric energy stored per unit volume is

$$W_{ave.} = \tfrac{1}{2}\epsilon_o\epsilon' E^2 \; (J/m^3) \qquad (16)$$

The second term at the right side of Equation 14 is the component of the current that is in phase with the electric field. Through this term, the microwave energy is converted into heat energy for material processing. The average power per unit volume converted into heat is,

$$P_{ave.} = \tfrac{1}{2}\omega\epsilon_0\epsilon' \tan\delta\, E^2 \;(W/m^3) \qquad (17)$$

Hence, the loss tangent characterizes the ability of the material to convert absorbed microwave power into heat with absorption depending on electric field intensity, frequency, loss factor, and permittivity. A "lossy" material (high $\tan\delta$ and ϵ'') heats more effectively than a low-loss (low $\tan\delta$ and ϵ'') material.

Conductive Losses

Electronic conduction can play a key role in the microwave heating of metal-like materials or semiconductors. As shown in Table 2-2, materials with moderate conductivity heat more effectively-than either insulating or highly conductive materials. Low-loss insulators are difficult to heat from room temperature, even though microwave penetration is significant. However, many oxide ceramics have resistivities that decrease rapidly with increasing temperature affording more efficient coupling (Newnham et al., 1991). As discussed later in this chapter and in Chapter 5, the rapid change in loss can lead to uneven heating and thermal runaway. Electronic conductivity does not vary significantly with the frequency in the microwave frequency range. The temperature dependence of the conductivity varies widely with the material depending on the dominant transport mechanism.

When the conductivity of the material is very large, the fields attenuate rapidly toward the interior of the sample due to skin effect. The skin effect involves the magnetic properties of the material. When a large current flows inside the sample due to a high conductivity, a combination of the magnetic field with the current produces a force that pushes conducting electrons outward into a narrow area adjacent to the boundary. The extent of this skin-arm flow is called the *skin depth, d_s*. Skin depth is defined as the distance into the sample at which the electric field strength is reduced to 1/e (Risman, 1991). The derivation is available in standard text books (Ramo and Whinnery, 1944; Iskander, 1992).

$$d_s = \frac{1}{(\tfrac{1}{2}\omega\mu_0\mu'\sigma)^{\tfrac{1}{2}}} \;(m) \qquad (18)$$

Skin depth ranges from several microns to a few meters. For example, at 2.45 GHz, brass and graphite have skin depths of 2.6 and 38 µm respectively and cured epoxy and alumina have skin depths of 0.73 and 187 m, respectively. When the skin depth is larger than the dimension of the sample, the effect may be neglected. In the opposite case, penetration of microwave energy will be limited, making uniform heating impossible.

Even though ions are thousands of times heavier than electrons and are chemically bonded into the network, ionic conduction losses are important in materials such as silicate

Table 2-2 Heating characteristics of a range of compounds (Newnham et al., 1991; Walkiewicz et al., 1988)

Material	Resistivity (Ω;-m)	Heating Characteristic (°C/min-temperature reached in specified time at 1 kW and 2.45 GHz)
Metal Powders	10^{-8} - 10^{-6}	**Moderately Heated**
Al		577°C/6 min
Co		697/3
Cu		228/7
Fe		768/7
Mg		120/7
Mo		660/4
Sulfide Semiconductors	10^{-5} - 10^{-3}	**Easily Heated**
FeS_2		1019/6.75
PbS		956/7
$CuFeS_2$		920/1
Mixed Valent Oxides	10^{-4} - 10^{-2}	**Easily Heated**
Fe_3O_4		1258/2.75
CuO		1012/6.25
Co_2O_3		1290/3
NiO		1305/6.25
Carbon and Graphite	~10	**Easily Heated**
Alkali Halides	10^4 10^{-5}	**Very Little Heating**
KCl		31/1
KBr		46/.25
NaCl		83/7
NaBr		40/4
LiCl		35/0.5
Oxides	10^4 - 10^{14}	**Very Little Heating**
SiO_2		79/7
Al_2O_3		78/4.5
$KAlSi_3O_8$		67/7
$CaCO_3$		74/4.25

glasses (Newnham et al., 1991). At low frequencies, ions move by jumping between vacant sites or interstitial positions in the network, giving rise to space charge effects. At higher frequencies, vibration losses, such as those from vibration of alkali ions in a silicate lattice, become important. Ionic conductivity does not vary much with the microwave frequency. Because ionic mobility is an activated process, the conductivity increases rapidly with temperature.

Atomic and Ionic Polarizations

Under an electric field, E, the electron cloud in an atom may be displaced relative to the nucleus, leaving an uncompensated charge-q at one side of the atom and +q at the other side. The uncompensated charges produce an electric dipole moment and the sum of those dipoles over a unit volume is the *polarization*, P. Similarly, deformation of a charged ion relative to other ions produces dipole moments in the molecule. Those are the atomic and ionic polarizations induced by the electric field. Although atomic and ionic polarizations occur at microwave frequencies, they do not contribute to microwave heating.

In terms of the polarization, ϵ may be written as:

$$\varepsilon = \varepsilon' - i\varepsilon'' = 1 + \frac{P}{E} = 1 + \frac{(P' - iP'')}{E}$$

Hence,

$$\varepsilon' = 1 + \frac{P'}{E}$$

$$\varepsilon'' = P''/E \qquad (19)$$

Therefore, in principle, the polarization contributes to both ϵ' and ϵ''. However, as shown schematically in Figure 2-13, atomic and ionic polarization mechanisms are active to optical and infrared frequencies, respectively. They act so fast that the net polarization observed under an electric field at microwave frequencies is in phase with the field. As a result, both P'' and ϵ'' are zero, and thus, atomic and ionic polarizations do not generally contribute to microwave absorption.

Losses associated with lattice or molecular vibrations in the infrared region due to the interaction of microwaves with dipoles are observed in alkali halides and in some polymer and composite systems (Newnham et al., 1991). These frequency shifts, or lower-frequency tails, are due to weaker bonding and heavier masses of heavy ions in the alkali halides, and to long-chain vibrations and weak interchain bonding forces in polymers.

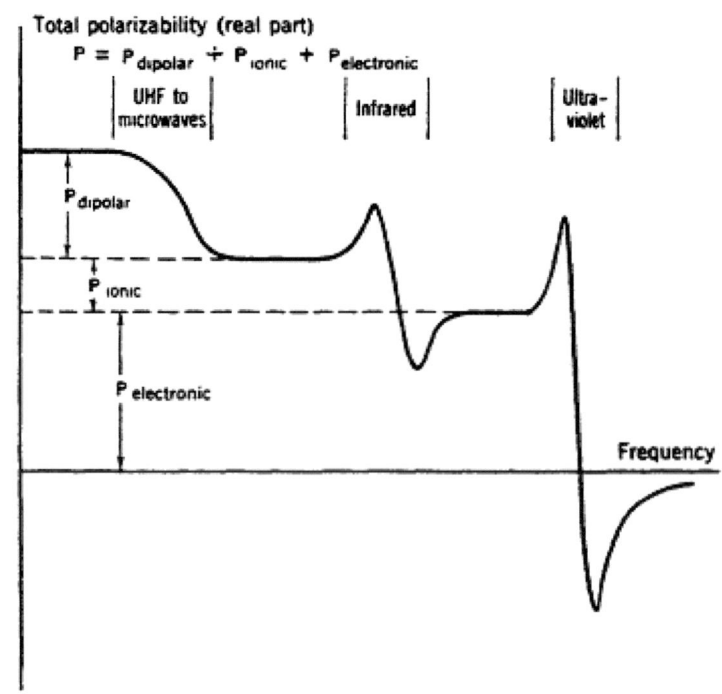

FIGURE 2-13
Frequency dependence of the several contributions to the polarizability schematic (Kittel, 1959).

Dielectric Materials and Electric Polarizability

In electronic conduction, either free motion of the electrons or collective diffusion of the ions is assumed. As the charge particles move, a current is induced. The situation is quite different in dielectric materials. Instead of the motion of the electrons or ions, electric dipoles play a dominant role in the properties of the material.

The mechanism of polarizability that causes the microwave absorption involves rotation and orientation of the dipoles (Kittel, 1959; Debye, 1929). There are three ways this can happen in a solid:

- The single atom can have the shape of the "electron cloud" surrounding the nucleus distorted by the electric field. In general, atoms with many electrons (high atomic number) are more easily distorted and are considered more "polarizable."
- Molecules with a permanent electrical dipole will have the dipole aligned in response to the electric field.
- Molecules, with or without a permanent dipole, can have bonds distorted (direction and length) in response to an electric field.

Without an electric field, the dipoles are randomly oriented and the net polarization is zero. In a static field, the dipoles align with the field and the polarization is maximized. These dipoles can rotate, but they rotate against a friction force. As the frequency of the electric field increases, the rotation of the dipoles cannot follow, and the net polarization in the material is no longer in phase with the electric field. In that case, it follows from Equation 15 that the polarization contributes both ϵ' and ϵ''. In terms of Debye's dielectric relaxation theory (Debye, 1929):

$$P = \frac{P_0}{1 + i\omega\tau} \quad (20)$$

Substituting Equation 20 into Equation 19:

$$\varepsilon' - 1 = \frac{P_0}{E}\left(\frac{1}{1 + \omega^2\tau^2}\right)$$

$$\varepsilon'' = \frac{P_0}{E}\left(\frac{\omega\tau}{1 + \omega^2\tau^2}\right) \quad (21)$$

The quantities $(\epsilon'-1)/(P_0/E)$ and $\epsilon''/(P_0/E)$ are represented in Figure 2-14. The relaxation time, τ, is the time interval characterizing the restoration of a disturbed system to its equilibrium configuration after a microwave field has been applied. The behavior of τ determines the frequency and temperature dependences of ϵ' and ϵ'' and varies widely in liquids and in solids. In water at room temperature, τ is about 5×10^{-11} s. The peak of the absorption, that is, the peak of ϵ'', is thus at the microwave wavelength of 1 cm. The absorption band is very broad. An interesting feature of this model is that the absorption disappears both at very low and very high frequencies, which explains very well the properties of water (Kittel, 1959). For polymers, τ is on the order of 10^{-7} seconds (Hawley, 1992). The unique feature of this model is that the effect of the polarization in both ϵ' and ϵ'' diminishes altogether above the microwave frequency.

The broad range of possible material properties that can be effectively processed using microwaves is illustrated by Table 2-3, showing representative dielectric properties of a sampling of important materials. Data needed for microwave process control or numerical simulations would require much more extensive information about the effect of temperature, frequency, and physical characteristics on the properties, as well as knowledge about heat-transfer responses. Although some excellent reviews exist for some materials (Bur, 1985; Westphal and Iglesias, 1971; Westphal and Sils, 1972; Westphal, 1975; 1977; 1980), there is a paucity of available data for both processors and heating-system designers.

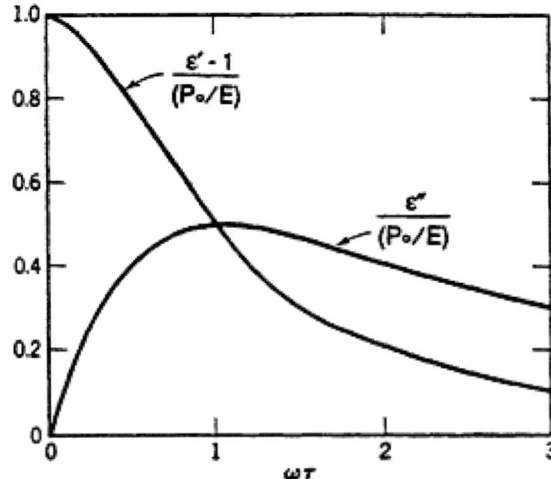

FIGURE 2-14
Frequency dependence of real and imaginary parts of the dielectric constant in the Polarization-Orientational Model (Kittel, 1959).

TABLE 2-3 Representative Dielectric Properties of Selected Materials.

Material	Frequency (GHz)	ϵ'	tan δ	Temp. (C)	Reference
Raw Beef	3.0	48.3	0.28	20	(Thuery, 1992)
Frozen Beef	2.45	4.4	0.12	-20	
Potato (78% water)	3.0	8.1	0.38	25	
Al_2O_3	3.6-3.8	9.02	0.00076	25	(Westphal and Iglesiad, 1971)
		9.69	0.00128	500	
		10.00	0.00930	700	
BN	8.52	4.37	0.00300	25	
Si_3N_4	8.52	5.54	0.00360	25	
Polyester	8.5	3.12	0.0028	25	(Westphal and Iglesias, 1971)
PTFE (Teflon)	2.43	2.02	0.00042	25	(Andrade et al., 1992)
PEI (Ultem)	1.0	3.05	0.003	25	(Bur, 1985)
Epoxy	1.0	3	0.015	25	(Bur, 1985)
Concrete (dry)	1.0	6.57	0.530	25	(Westphal and Iglesias, 1971)
Concrete (wet)	1.0	13.2	0.485	25	

Depolarization Factors

In a dielectric material, as the sample is polarized under an applied electric field, surface charges appear at the sample boundary (Figure 2-15). The surface charges produce a depolarization field opposing the applied field. The net field inside the sample is thus reduced, while the field outside the sample remains the same. The computation in this case involves a depolarization factor, N, listed in Table 2-4, which is simple only if the sample has an ellipsoidal shape and has its major axis placed either parallel or normal to the electric field. Fortunately, a cylinder may be considered an elongated ellipsoid and a disc, a fiat ellipsoid. Taking the depolarization factor, N, into account, the relation between the field inside the sample, E_{in}, and that outside, E_{ext}, is (Kittel, 1959; Becker, 1964):

$$E_{in} = [1 - \frac{N(\varepsilon'-1)}{1+N(\varepsilon'-1)}]E_{ext} \qquad (22)$$

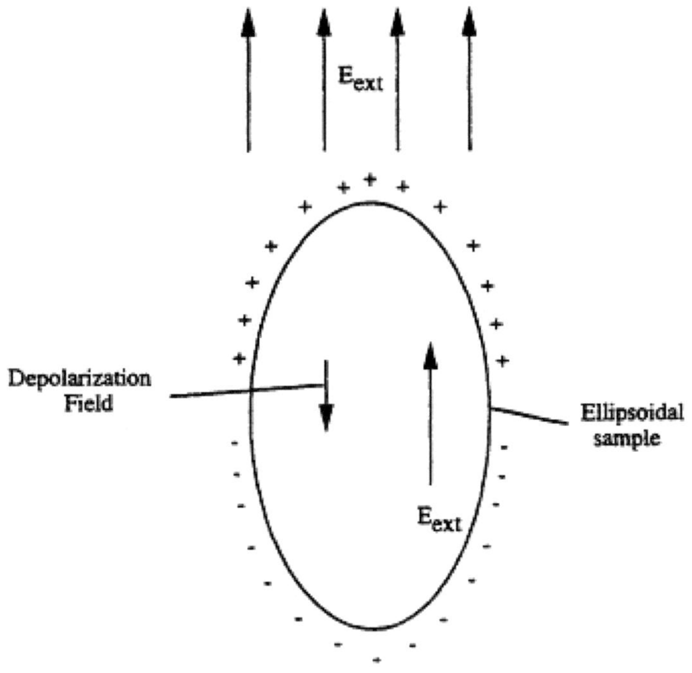

FIGURE 2-15
The field outside the sample E_{ext} and that inside is the sum of E_{ext} and a depolarization field (adapted from Kittel, 1959).

When $N\epsilon'$ is large, the electric field inside the sample could be reduced to zero. The sample is then completely shielded from effects of the microwaves. To avoid this, surface charges must be reduced. This could be accomplished by using a slender cylinder or a thin disc placed parallel to the electric field (N=0). Due to either skin effect or depolarizing fields, microwave heating of bulk materials that have very large conductivity or permittivity is very difficult.

TABLE 2-4 Depolarization Factors, N

Shape of the sample	Depolarization Factor, N
Sphere	1/3
Thin slab normal to the field	1
Thin slab parallel to the field	0
Long cylinder parallel to the field	0
Long cylinder normal to the field	1

Thermal Runaway

The discussion of microwave fundamentals has focused on the generation and propagation of microwaves and their interaction with materials. However, the thermophysical behavior of the sample must also be understood. Stable microwave heating depends on the rate of microwave power absorption and on the ability of the sample to dissipate the resulting heat, that is, if the temperature dependence of the power absorption is less than the temperature dependence of the heat dissipation at the surface of the specimen (plus insulation system), stable heating should be observed. The rapid rise in dielectric loss factor with temperature is the major issue in thermal runaway and temperature nonuniformity. Therefore, although microwave heating frequently is touted as providing more uniform heating, nonuniform heating is a reality in many materials, often at nominal heating rates.

Some attempts have been made to quantify the conditions under which thermal runaway occurs and how it can be controlled. Roussy et al. (1985; 1987) predicted that heating rubber specimens above a certain power level was unstable, and that temperature increased uncontrollably, while below that power level the temperature came to a steady state. The regime in which stable heating occurred was mapped in terms of the heat loss and acceleration of power absorption with temperature rise. Stable heating was possible with rapid heat removal (i.e., no insulation) or small temperature dependence of the dielectric loss factor. Tian et al. (1992) used computer simulation of two-dimensional temperature distributions in microwave-heated alumina ceramics to predict that up to certain power levels, stable heating in both uninsulated and insulated specimens would occur. As expected, the critical power levels were greater in uninsulated specimens. The use of insulation significantly accelerated both the heating rate and the risk of thermal runaway.

Kriegsmann (1992) modeled heating of uninsulated ceramic slabs and cylinders. Taking account of the effect of material properties on the microwave field within the materials, but ignoring temperature gradients, it was determined that, over a certain incident power range, the part temperature is a multivalued function of incident power resulting in an S-shaped power-temperature response curve (Figure 2-16). Below the critical power level, the sample will heat in a stable manner to a steady-state value on the lower branch of the response curve. If the power is increased to exceed the upper critical power, the temperature will jump to the upper branch of the temperature—power curve. These observations have been supported by modeling work performed to simulate microwave heating of alumina (Barmatz and Jackson, 1992; Johnson et al., 1993). These studies emphasize the importance of sample size, geometry, relative density, and composition.

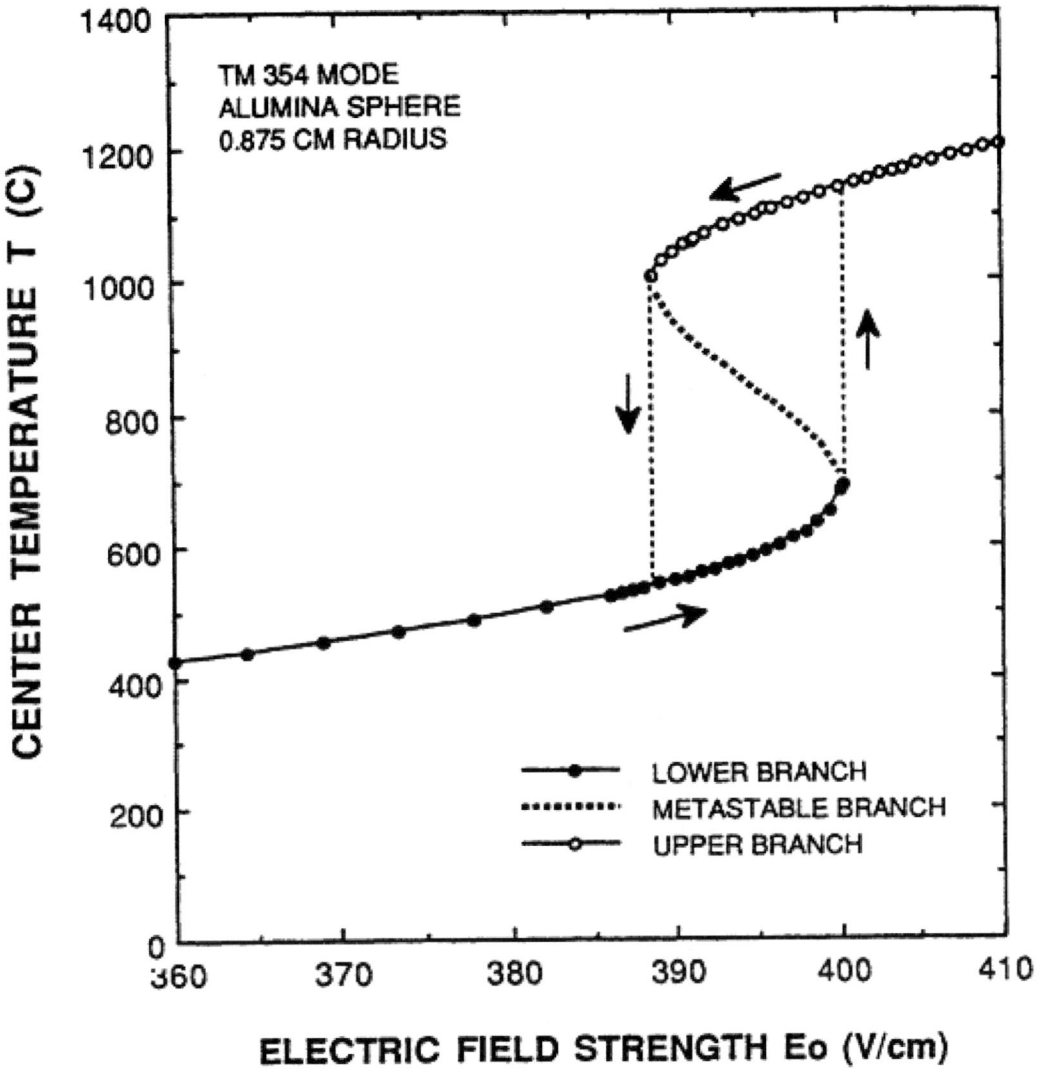

FIGURE 2-16
Predicted power——temperature response curve for an alumina sphere (courtesy of M. Barmatz, Jet Propulsion Laboratory).

Due to rapidly increasing dielectric loss factor, the area of the sample that first exceeds the critical temperature will continue to heat rapidly at the exclusion of the rest of the sample. Thus, process control schemes to control thermal runaway depend upon knowing the temperature at the interior of the specimen. While it may be that precision is more important than accuracy for this control, many of the problems of temperature measurement discussed in Chapter 3 will influence the processor's ability to control the process. Moreover, in industrial practice, it is not usually possible to insert a thermocouple or optical probe into the specimen.

3

Microwave System Integration

A considerable investment has been made over many years in the development of microwave processing systems for a wide range of products including food, rubber, ceramics, and a vast array of other highly specialized products. Most of these have been developed on a "bootstrap" basis by small, industrial microwave companies working with users in joint development arrangements in which the user gets a proprietary process and the microwave company gets an exclusive manufacturing agreement if the process succeeds. As a result, much of the technology that has been developed to date is not widely available, and the "wheel" has to get reinvented over and over again as others explore the potential benefits of microwave processing.

While the basic components of a microwave processing system——generator, applicator, and control systems——are simple, the interaction of materials with microwave fields and changes in fundamental material properties during processing make design and development of microwave processes very complex. This complexity may be dealt with using an integrated approach with a process design team consisting of the materials and process engineer, the microwave equipment manufacturer, and an electromagnetic specialist.

This chapter identifies key considerations in equipment selection and design, consistent measurement of sample temperature in a microwave field, measurement of critical properties, and numerical process simulation. Use of these tools and an interdisciplinary approach to system design will lead to more successful applications of microwave technology.

MICROWAVE APPLICATORS

Simply stated, microwave applicators are devices that are designed to heat a material by exposing it to a microwave field in a controlled environment. The objective is to cause a controlled interaction between the microwave energy and the material to occur under safe, reliable, repeatable, and economical operating conditions. Applicators may be conveyor operated; batch; or, as in the case of indexing systems, a combination of both. Microwave energy may also be combined inside the applicator with other energy sources, such as hot air, infrared, and steam, in order to achieve special results. Microwave applicators may also be designed to permit controlled interaction under a variety of ambient conditions, ranging from vacuum to high pressure and humidity.

Multimode Applicators

As described in Chapter 2, the two general classes of microwave applicators are multimode and single mode cavities. Key features of multimode ovens include:

- suitability for bulk processing applications;
- oven dimensions that are often determined by product dimensions;
- moderate to high efficiency;
- adaptability to batch or continuous product flow;
- performance that is less sensitive to product position or geometry; and
- good uniformity that may require motion of product or hybrid heating.

Multimode applicators are often used for processing bulk materials or arrays of discrete material, whose overall dimensions are too large (larger than the wavelength of the operating frequency) to permit consideration for use in a single-mode oven. These applicators, in their simplest configuration, take the form of a metal box that is excited (driven) at a frequency well above its fundamental cutoff frequency. For example, the common home microwave oven typically has internal dimensions on the order of 12 to 16 in., while the wavelength is 4.8 in. The larger dimension corresponds to a cutoff frequency of about 400 MHz as compared with the operating frequency of 2.450 GHz.

Because the dimensions of the enclosure are very large when expressed in terms of the free-space wavelength of the operating frequency, a large number of standing-wave modes can exist at or very near the operating frequency inside the cavity. To establish a reasonably uniform electric field strength throughout the cavity, it is desirable to excite as many of these modes as possible. When multiple modes are excited, heating nonuniformity is minimized even when the field perturbing effects of the materials being processed are present.

Multimode applicator design involves a number of basic design parameters. They include uniformity of heating, required microwave power, applicator size, leakage suppression, and required performance characteristics.

Heating Uniformity

Uniform heating is difficult to obtain in a multimode oven. This difficulty arises from the unpredictable way in which the parameters affecting uniformity change with time. As a result, a number of techniques, in addition to excitation of multiple standing-wave modes, are used to promote uniform heating. They include metallic mode stirrers to ensure that all the possible modes are excited; surface scanning to direct the energy at regions of interest; product motion; and, in some cases, hybrid heating using conventional heating to replace surface losses.

In conveyor applicators, product motion is inherent in the process. In batch applicators, product motion may be introduced in a variety of ways that include rotation, orbital motion, and linear (vertical or horizontal) translation.

Another approach to improving heating uniformity is evident in the recently developed variable frequency microwave furnace. This furnace is a multimode cavity driven by high-power traveling wave tubes that produce up to 250 W continuous wave from 2.5 to 17.5 GHz and 500 W continuous wave from 0.9 to 8 GHz (Lauf et al., 1993). Earlier work showed capability to generate up to 2.5 kW from 4 to 8 GHz (Bible et al., 1992). As shown in Figure 3-1, the variable frequency microwave furnace consists of a traveling-wave-tube amplifier capable of sweeping approximately an octave frequency range and a signal control system that can adjust frequency or power to maximize absorption. Samples can be heated at fixed frequency, with adjustments for changes in sample characteristics, or subjected to continuously swept frequencies to achieve large-area uniformity by sequentially establishing several cavity modes.

Required Microwave Power

Required power is usually calculated based on an initial assessment of the proposed process and verified through actual testing once an initial oven concept, layout, and size have been established. Key parameters to be verified include heating-rate sensitivity, temperature uniformity, and process efficiency. Rate sensitivity, which can be a problem in some drying applications, may force the use of a longer cavity to increase process time at the expense of process efficiency.

Applicator Size

In many applications, applicator size is determined largely by the product size and compatibility with existing factory conveyor or batch production formats. Product size and rate sensitivity issues, mode number (uniformity requirement), and microwave-power handling capability under no-load conditions usually determine the minimum size of an oven. It is essential that an applicator be capable of operating under no-load conditions without electric field breakdown and without leakage for at least a sufficient time to let equipment and personnel safety devices shut the system down. Conveyorized ovens with large entrance and exit tunnels usually employ "lossy" (high ϵ'') walls to suppress leakage. Under no-load conditions, the lossy walls act as parasitic loads that help reduce field strength in the cavity, thus reducing the risk of destructive arcing.

Many batch ovens are designed to process a "disappearing load," which means that at the end of the process the cavity is effectively empty. A good example of this type of application is the drying of refractory materials. The microwave energy couples primarily to the water, since the refractory is relatively transparent to the microwaves. As drying proceeds, the amount of moisture in the cavity, and thus the microwave load, decreases. When no-load conditions are reached, extremely high fields can exist in the cavity and in waveguides feeding the cavity. Under these conditions, destructive arcing is a possibility unless special precautions have been taken in the design m prevent it. Applicator design must allow for this possibility by allowing no-load operation or by providing means (e.g., arc detectors) to automatically shut down when safe operating conditions are exceeded.

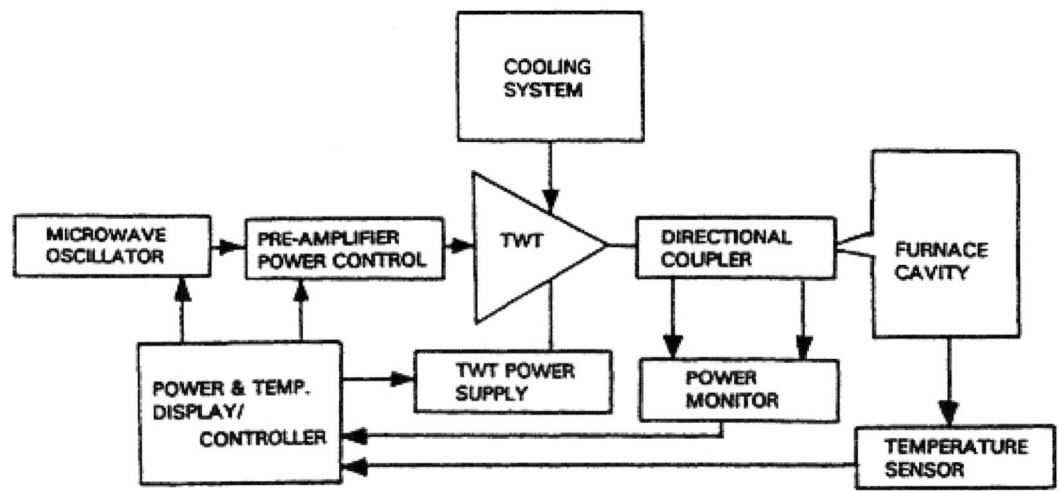

Figure 3-1
Schematic diagram of variable frequency microwave furnace (Courtesy of Microwave Laboratories, Inc.).

Leakage Suppression

Suppression of microwave leakage from microwave oven doors and product openings is required for personnel safety and to reduce electromagnetic interference. Although these are two very different issues, they must be dealt with simultaneously by one choke or suppression runnel design. The current safety standard for microwave ovens is an emission specification that limits emissions at a distance of 5 cm from the surface of an oven to a maximum of 5 mW/cm^2. Safety standards are discussed in more detail in a later section of this chapter.

Leakage can usually be suppressed by means of reactive chokes, provided that the other dimension of the opening is less than approximately one-half of a wavelength. Good examples of these types of openings are the door seals for industrial and conventional home microwave ovens and slot openings to permit ingress and egress of thin belt wed web materials processed in industrial microwave ovens.

Reactive chokes are ineffective when the height of the opening is grater than about half a wavelength. In these cases, free radiation from the cavity can occur with the possibility of unacceptable levels of human exposure. There are three basic methods employed to deal with these situations. They are (1) leakage suppression runnels with absorbent walls, (2) vestibules with indexing conveyors and doors that open and close sequentially to admit product, and (3) "maze" openings that admit product by causing it to meander through a folded corridor lined with absorbing walls. These are illustrated schematically in Figure 3-2.

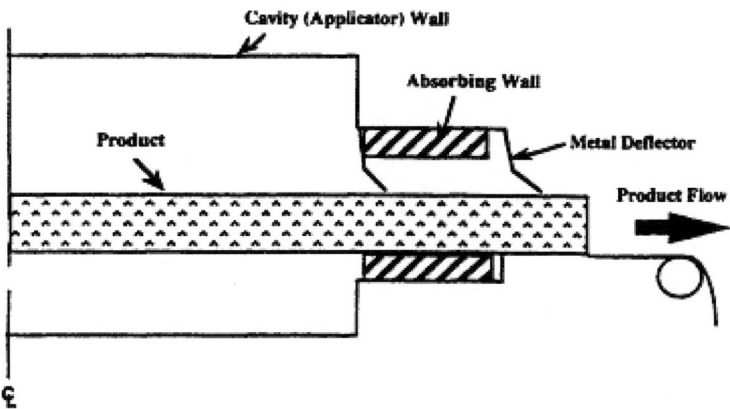

(a) Lossy Wall Leakage Suppression

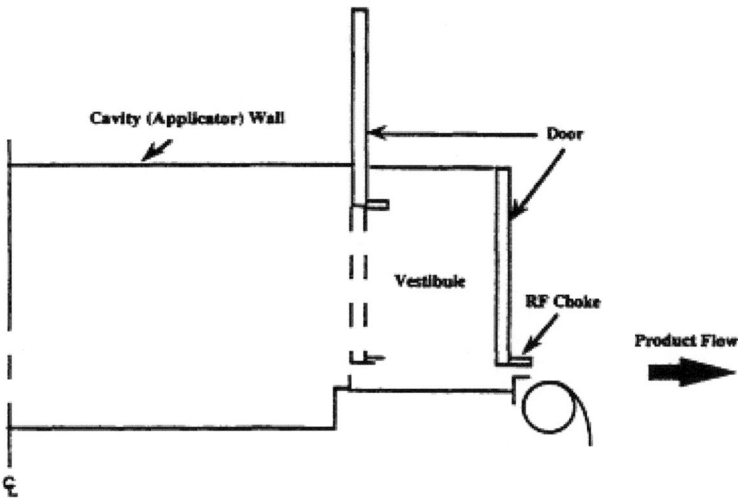

(b) Vestibule Leakage Suppression

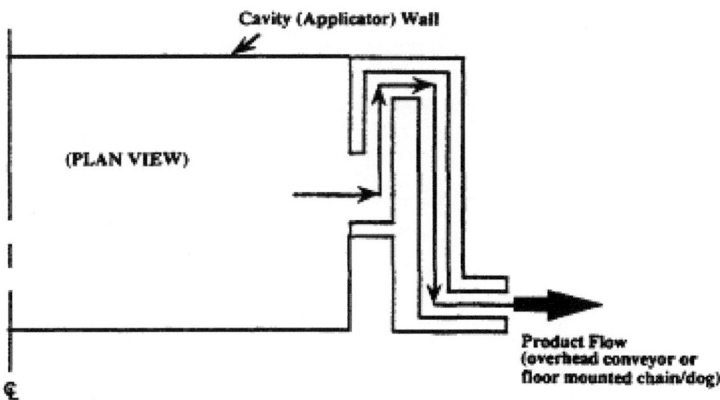

(c) "Maze" Leakage Suppression

FIGURE 3-2
Schematic of leakage suppression approaches.

Required Performance Characteristics

Figure 3-3 illustrates typical temperature versus heating time curves for most of the products processed in a multimode oven under constant microwave power input. Generally speaking, the slope of the heating curves, at low temperatures at least, will be linear provided that the change in material dielectric properties and heat loss mechanisms in the range of interest are negligible. The low-loss case (curve C) is typical for material such as natural rubber and other polymers that have low dielectric loss. Curve B is representative of a moderate-loss material or a low-loss material (curve C) doped to increase its loss. The plateau in curve B could represent the thermostating effect of a drying process that occurs when the moisture reaches the saturation temperature. The tail end of curve B represents what happens after all of the moisture is gone. If the base material is low-loss or transparent, the material will stop heating, and the temperature will gradually decrease. If the material is lossy, the heating process will continue upward until heat losses from the surface (radiative, conductive, and convective) occur at the same rate of the microwave energy is being delivered to the oven. This last effect is illustrated by curve A.

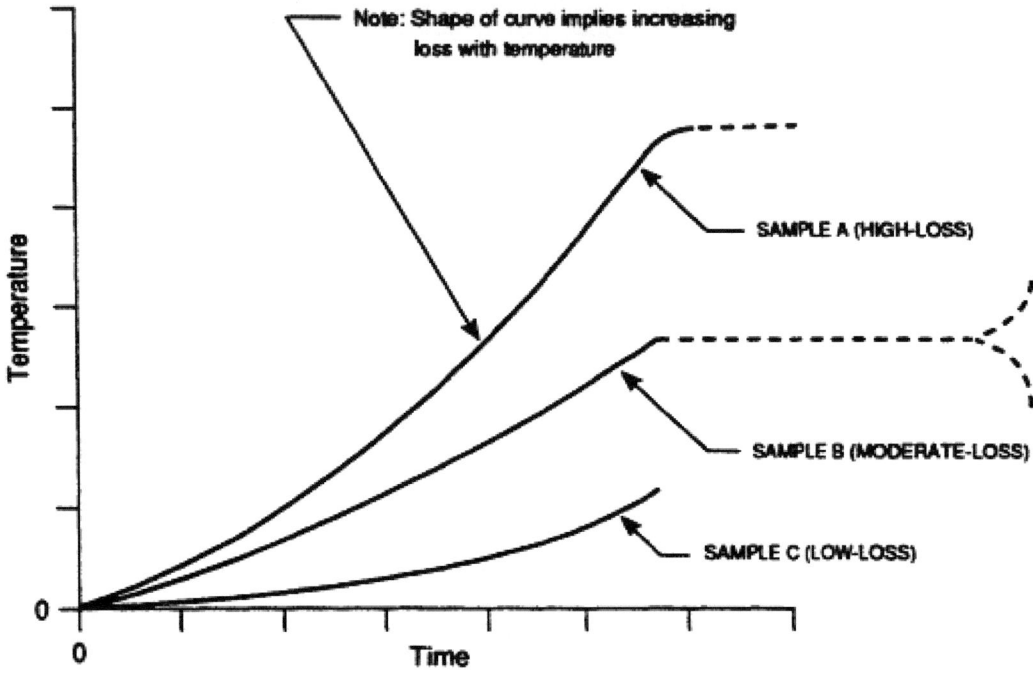

FIGURE 3-3
Typical temperature vs. heating time curves for multimode applicator with constant microwave power, sample weight and sample geometry.

Figure 3-4 illustrates how the efficiency of a multimode cavity is affected by the amount of material in the cavity, and the dielectric loss properties of the material.

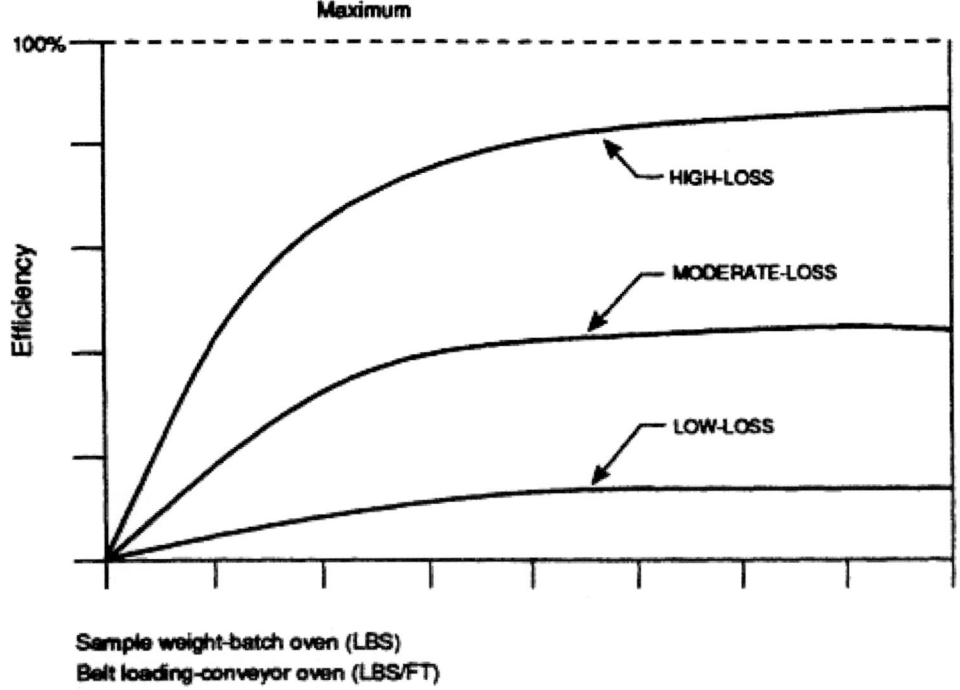

FIGURE 3-4
Schematic of multimode cavity efficiency vs. load size with constant power and variable sample weight.

Single-Mode Applicators

In their simplest form, single-mode applicators consist of a section of waveguide operating at a frequency near cutoff. They usually have had holes or slots cut in them to let product in or out. In more-demanding applications, they may consist of resonant, high Q cavities. Some advantages of single-mode applicators follow:

- High fields are possible.
- The applicators can operate in the standing or traveling wave configurations.
- Fields are well defined.
- Fields can be matched to product geometry.
- The applicators are useful for heating both low-loss and high-loss materials.
- The applicators are compatible with continuous product flow.
- High efficiency is possible.

The use of single-mode applicators involves some penalties that must be weighed carefully. They are product specific rather than general purpose and in operation can be very sensitive (i.e., tuned off-resonance) to changes in product properties, geometry, and position.

As a result of these shortcomings, single-mode applicators tend to be more expensive to design. Their use may require automatic controls and feedback to ensure optimum performance.

Figure 3-5 illustrates two examples of single-mode applicators. The first is a single-mode waveguide with a slot cut into the broad wall to permit passage of a thin material that is to be heated. Cutting the slot in the center of the broad wall of the cavity where the TE_{10} electric field is at a maximum ensures efficient coupling. The second is a cylindrical cavity operating in the TM_{01} mode, which puts the E-fields parallel to the longitudinal axis of the cylinder.

Periodic Structures

Periodic structures are microwave transmission lines that have been distorted or had discontinuities inserted into them at regular intervals. When microwave energy is fed through these structures, there are reflections caused by the discontinuities which result in the formation of pass bands and stop bands. These structures are important because field shaping to match product geometry is possible and because single-surface nonradiating devices can be built. As with single-mode applicators, periodic applicator design is involved and expensive, and the final design is very product specific.

Two examples of these structures are shown in Figure 3-6. The first is the serpentine wave guide, which is similar to the slotted wave guide (Figure 3-5a) except that the structure has been folded back on itself to accommodate wide materials and to provide uniform heating and sufficient interaction length to ensure absorption of all of the microwave energy. The second is a single-surface applicator (Karp Line) formed by cutting slots in the broad wall of a rectangular wave guide. Microwave fields fringe out to the external surface, where they interact with a thin-film material to be heated. This structure has been used successfully to cure the glue used to bind the signatures in a book binding application.

MICROWAVE SAFETY STANDARDS

As microwave power levels for industrial processing systems increase, potential hazards associated with exposure to radiation become more important. Extensive work, as summarized in recent review articles (Michaelson and Lin, 1987; CRC, 1986), indicate that the effects on biological tissue from exposure at microwave frequencies are thermal in nature. Unlike the higher energy, ionizing region of the electromagnetic spectrum, including x rays and gamma rays, the nonionizing bands from DC to visible light do not carry enough energy to break chemical bonds (Redhead, 1992). The only effects of nonionizing radiation in the microwave region on human tissue are those derived from the energy—matter coupling mechanisms, particularly dielectric coupling, described in Chapter 2. At present, the only confirmed effect is warming, from the conversion of electromagnetic energy to heat. Thus, microwave exposure standards are based on the thermal effects of exposure.

Since, at microwave frequencies, sensation and pain thresholds are well before burns occur (Osepchuk, 1991), burns are most likely to occur due to contact with a heated conductor or opening, rather than through radiant exposure.

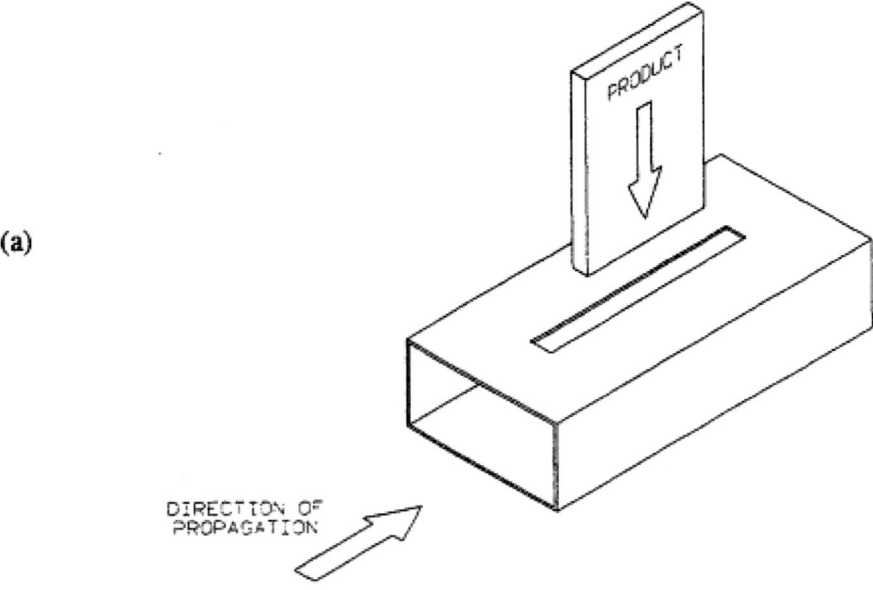

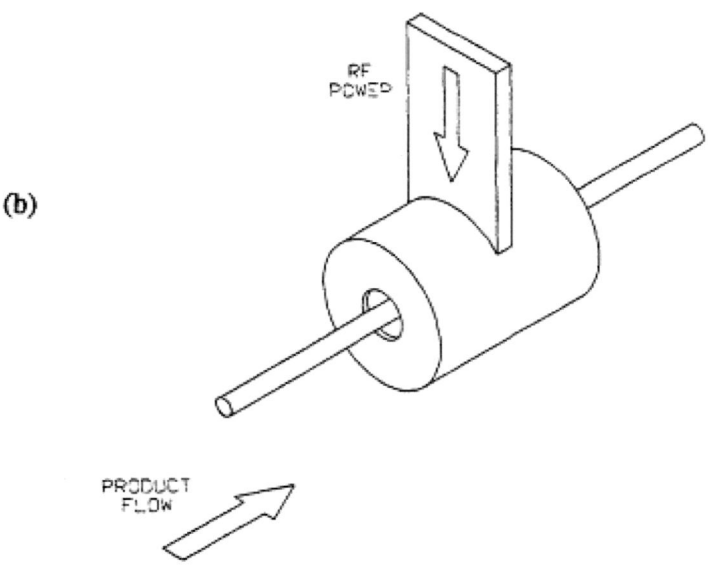

FIGURE 3-5
Single-mode applicators: (a) single-mode waveguide; (b) cylindrical single-mode cavity.

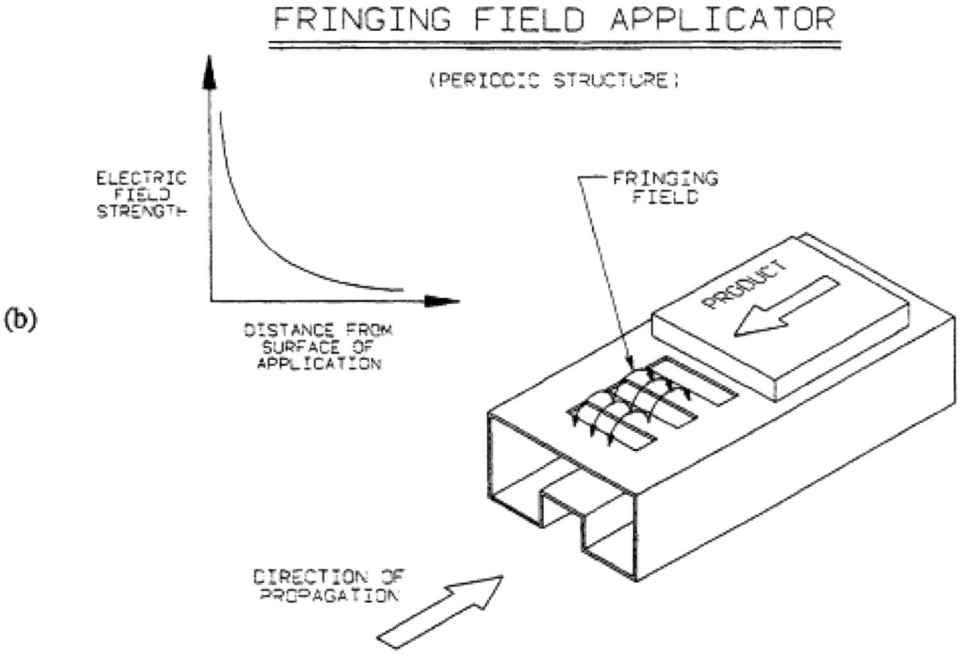

FIGURE 3-6
Periodic applicators: (a) serpentine waveguide; (b) fringing field.

The currently accepted standard is the guidelines developed by the American National Standards Institute of 10 mW/cm^2 power density for exposure (ANSI C95.1-1991). The power density guideline is based on a maximum permissible exposure of 0.4 W/kg specific absorption rate, which is a factor of 10 less severe than the determined threshold absorption level (Redhead, 1992). Standards based on the institute's guidelines include the Food and Drug Administration's emission standard of 5 mW/cm^2 at 5 cm for microwave ovens (HHS, 1991) and the Occupational Safety and Health Administration's exposure standard of 10 mW/cm^2 (Department of Labor, 1991).

To minimize exposure, the microwave system needs to be designed with effective leakage suppression, viewing or ventilation screens, and an interlock system on doors and access apertures to shut off power when doors are opened (Osepchuk, 1991).

TEMPERATURE MEASUREMENTS

One of the most difficult, yet important, parameters to measure in a microwave environment is temperature. Sample temperature is the most common process-control parameter in microwave processing. Inaccuracies in temperature measurement or perturbation of the microwave field by temperature sensors can lead to erroneous indications of process temperature and misleading representation of process efficiency.

The temperature of a body is its thermal state and is regarded as a measure of its ability to transfer heat to other bodies. The indication of how a numerical value may be associated with the temperature requires a review of the laws of thermodynamics, which is certainly beyond the scope of this document. To establish procedures for accurate temperature measurement, temperature may be defined as a quantity that takes the same value in two systems that are brought into *thermal contact* with one another and are allowed to come to *thermal equilibrium*. Based on this definition, it may be suggested that for accurate temperature measurements, both the body and the measuring device should make good thermal contact and both bodies should be in thermal equilibrium.

Figure 3-7 identifies various temperature-measuring instruments and their applicable temperature ranges. Thermocouples and optical measurement techniques are most often used in microwave processing.

Temperature Measurement in a Microwave Environment

Temperature measurements in a microwave environment present several difficulties:

- Temperature measurement must be made directly within the sample and not in its vicinity. Microwaves heat the sample itself (heat from within) and not the surroundings, and hence temperature probes must maintain good contact with the sample to achieve accurate temperature measurements.
- Thermal gradients developed during microwave heating make characterizing sample temperature using a single measurement difficult.

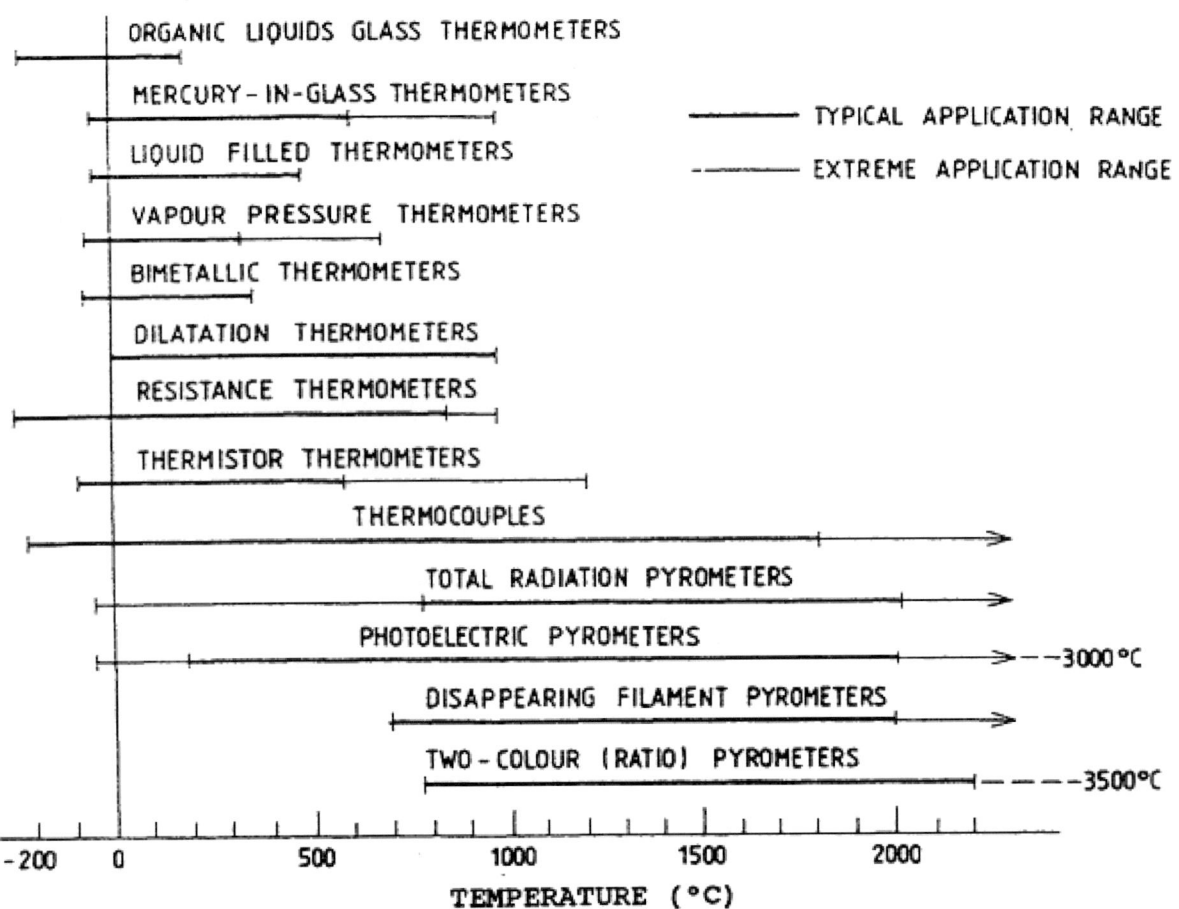

FIGURE 3-7
Classification of temperature-measuring instruments.

- Maintaining good contact with the sample might be difficult because of the changes in sample size during processing or due to motion of the sample.
- Conventional temperature-measurement procedures using thermocouples are not suitable for making these measurements in a microwave environment. The presence of a metallic temperature probe in a microwave environment can cause electromagnetic interference problems, causing distortion of the electrical field or affecting the electronics used for temperature measurement, as well as errors due to self heating; heat conduction; shielding; and excessive localized heating, particularly at the tip of the probe.
- Optical measurement techniques such as pyrometers and optical fiber probes assume knowledge of emissivity.
- The "heating from within" property of microwave heating results in a surface temperature that is different from the core value.
- Depending on the optical properties and size of the sample, its emittance may make the optical radiation sensitive to the colder environment surrounding the sample.

Some of these measurement errors will be discussed in more detail in the following section.

EMI Problems

When metallic temperature probes are used in the microwave environment, significant electromagnetic interference problems may occur. Electromagnetic boundary conditions require that the electric field be normal to conductors. Therefore, serious distortion of the electric field in the microwave environment will occur when a thermocouple is placed at a location where the electric field is parallel to the conductor. Figure 3-8a, shows results from numerical simulations, where it may be seen that placement of a conductor parallel to electric field vectors caused significant distortion in the field distribution in a multimode cavity. Figure 3-8b, on the other hand, shows that less distortion occurs when a metallic conductor is placed in a region of small electric field values, particularly since the conductor is perpendicular to the electric field vector. The only difference between parts (a) and (b) of Figure 3-8b is a noted concentration of the electric field lines at the tip of the conducting probe. This is the reason why it is often observed that using metallic temperature probes results in localized heating and possible thermal runaway in the sample.

Alumina sheathed thermocouples have been used to reduce the effect of localized heating (Janney et al., 1991a). Figure 3-9 shows that, based on the boundary condition of the normal electric field component, such an arrangement will result in minimizing the local concentration of the electric field near the tip of the probe. However, such an arrangement is not expected to minimize the distortion of the electric-field distribution in the cavity, particularly when the probe is placed parallel to the electric field.

Although it may be acceptable to insert metallic temperature probes in multimode microwave cavities, placement of such probes perpendicular to electric field lines is crucial when measuring temperature in a single-mode cavity. Fortunately, electric field distributions are known in cavities, and hence it is possible to make such adjustments in certain modes.

Otherwise, serious distortions in the field configuration are expected, and performing a heating process under these conditions might be meaningless.

Errors in Optical Temperature Procedures

Although supplier specifications of accuracy may be as good as 2 °C or less, there are many conditions that need to be satisfied before such accuracies may be achieved.

First, optical techniques measure surface temperature, which may be significantly different than internal temperatures due to microwave penetration and surface heat losses. Relationships between surface measurements and actual internal temperatures need to be understood.

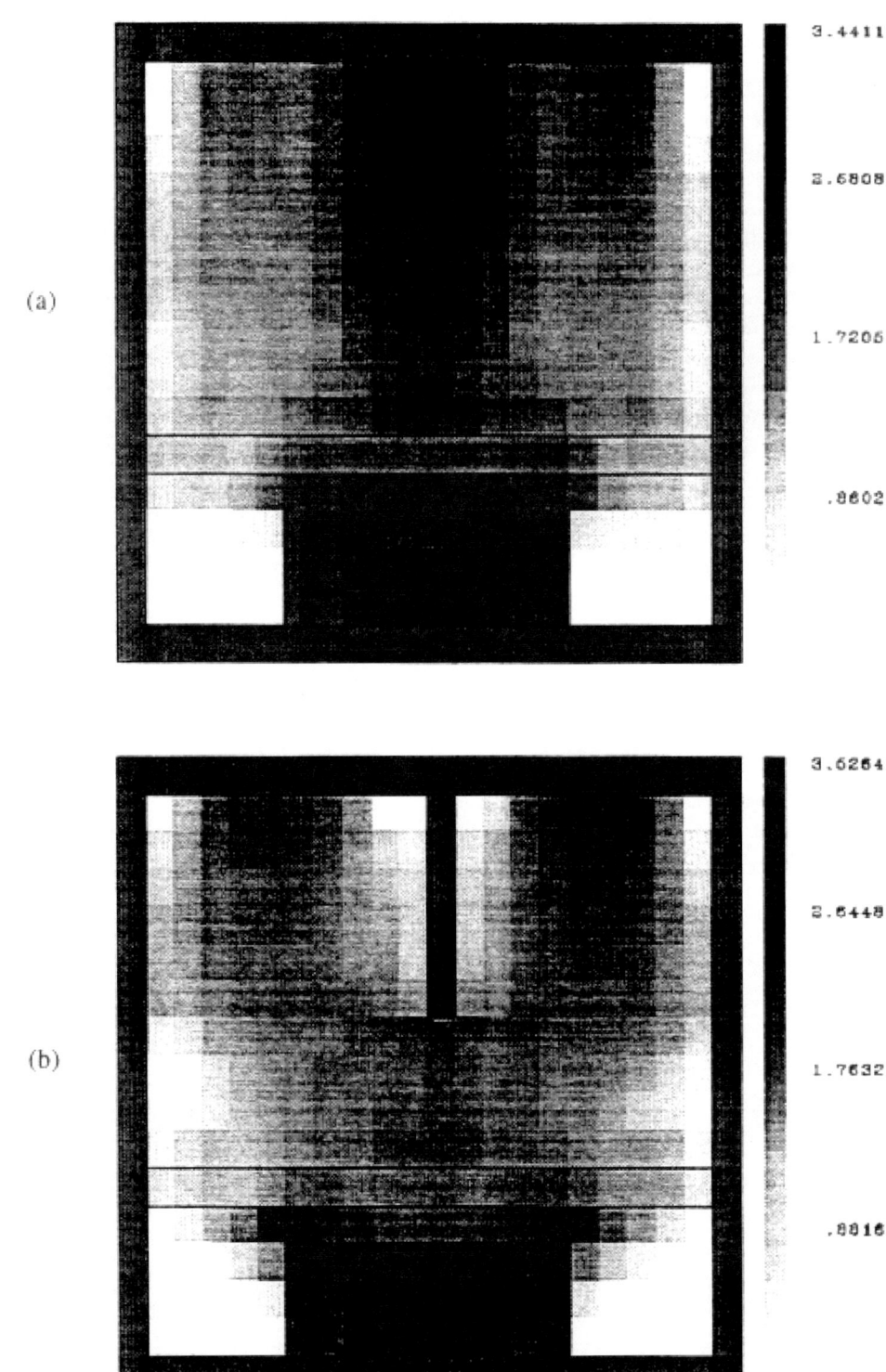

FIGURE 3-8a
Electric field intensity distribution in a vertical plane through a multimode cavity and a ceramic base plate. Plot (a) is the empty cavity, and plot (b) includes a thermocouple placed parallel to the majority of the electric fields.

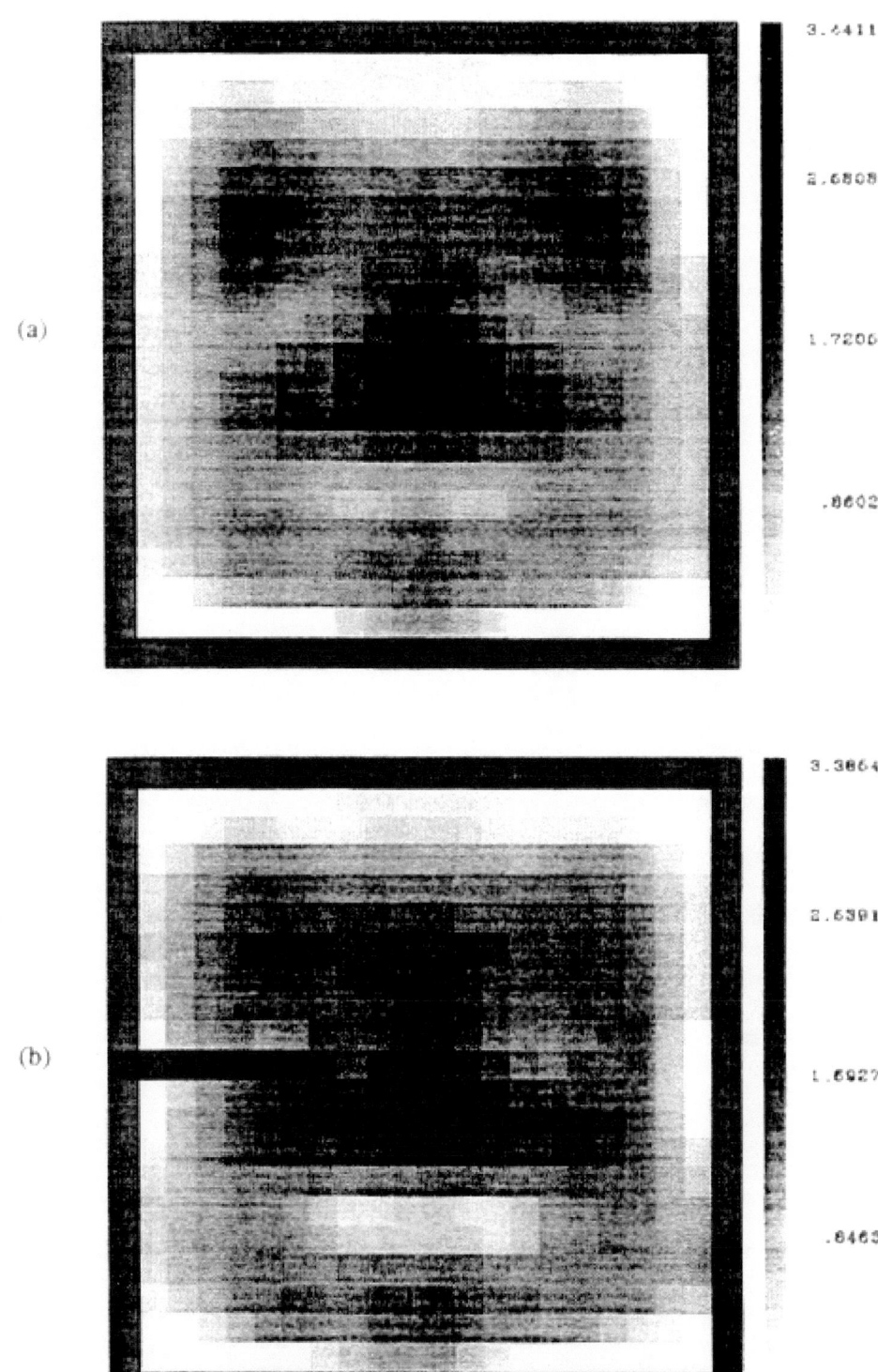

FIGURE 3-8b
Electric field intensity distribution in a horizontal plane through a multimode cavity above a ceramic base plate. Plot (a) is the empty cavity, and plot (b) includes a thermocouple place perpendicular to the majority of the electric fields.

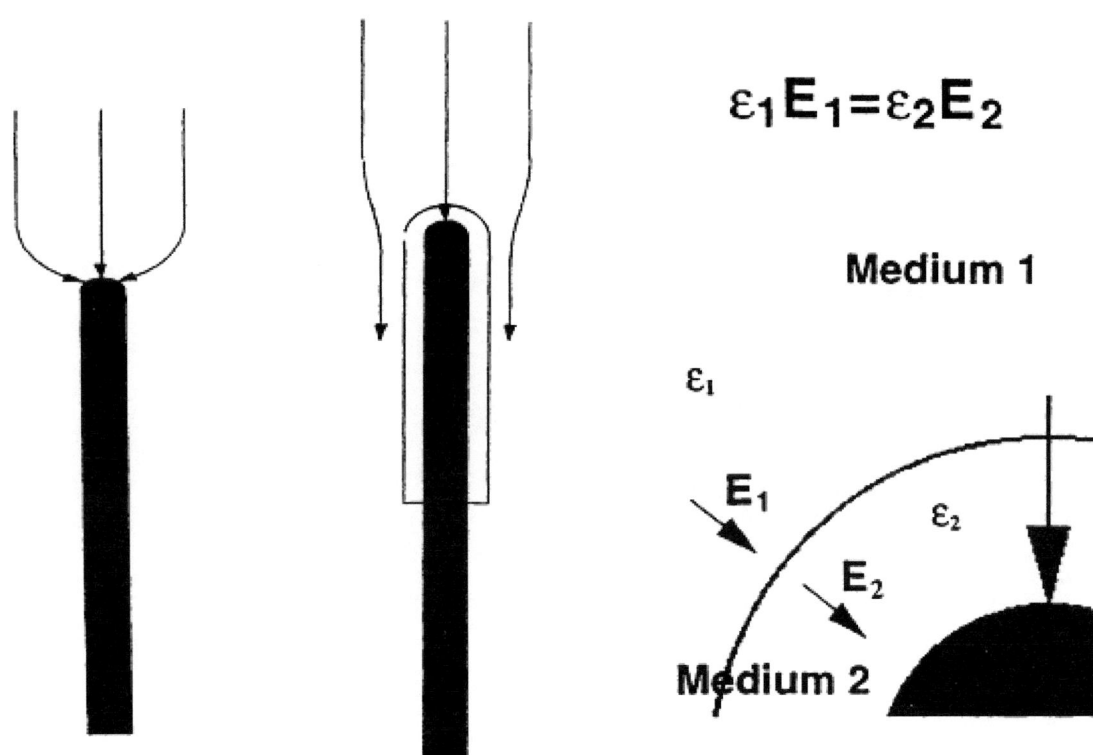

FIGURE 3-9
Effect of ceramic sheath on electric field concentration at temperature probe tip (Kimrey and Iskander, 1992).

Second, accurate knowledge of the object's emissivity and its variation with temperature and frequency is possibly the most serious source of error in both pyrometric and optical-fiber thermometer measurement techniques. Emissivity is the ratio between the radiation energy of an object and the energy radiated by a perfect "blackbody" radiator at the same temperature. Although emitted radiation from a blackbody may be quantified in terms of Planck's blackbody radiation law, accurate knowledge of the emissivity of an object and its variation with temperature and frequency requires careful analysis. Emissivity depends on the object's surface state, homogeneity, and temperature. The error in measured temperature, dT, is related to an error in the emissivity values $(\epsilon_n - \epsilon_o)$ by

$$dT = \frac{2(\epsilon_n - \epsilon_o)}{(\epsilon_n + \epsilon_o)C_2} \lambda T^2$$

where ϵ_n is the new emissivity value, ϵ_o is the original emissivity value, $C_2 = 1.4387 \times 10^4$ μm·K, and λ is the measurement wavelength in μm. Figure 3-10 shows the temperature-measurement error based on percent error in emissivity as a function of temperature. Figure 3-10, part a shows the case when temperature measurements were made at $\lambda = 0.95$ μm, while Figure 3-10, part b shows the expected errors when the measurements are made at $\lambda = 3.5$ μm, which is conventionally used in these measurements. Errors as large as 200 °C may occur if the emissivity estimate was in error by 20 to 25 percent at $\lambda = 3.5$ μm. Additional error may result from high sample transparency (emittance), which causes the temperature measurements using pyrometers or lightpipe to be sensitive to the cooler environment surrounding the heated sample.

Finally, pyrometer temperature measurement in a single-mode cavity is often made through a Pyrex window. It is therefore important that the effect of the window be carefully calibrated, and fogging effect (if any) should be taken into account. Figure 3-11 shows the spectral transmittance of a 12.7-mm Pyrex window. The fact that the transmittance of the window depends on frequency adds another source of error when a dual-wavelength pyrometer is used in temperature measurement. To help quantify this error, consider Planck's equation for the radiated energy from a blackbody as a function of temperature and wavelength.

$$E(\lambda,t) = \frac{C_1 \lambda^{-5}}{e^{C_2/\lambda T} - 1}$$

where λ is the wavelength, T is the temperature in Kelvins, $C_1 = 3.743 \times 10^8$ W·(μm)^4/m^2, and $C_2 = 1.4387 \times 10^4$ μm·K. If we consider a dual-wavelength pyrometer operating at 0.8 and 0.95 μm, then

$$\frac{E(0.80,T)}{E(0.95,T)} = \frac{\dfrac{C_1(0.8)^{-5}}{\left(e^{-\frac{C_2}{0.8T}} - 1\right)}}{\dfrac{C_1(0.95)^{-5}}{\left(e^{-\frac{C_2}{0.95T}} - 1\right)}}$$

$$T = \frac{14387}{\ln\left(\dfrac{E_{0.8}(0.8)^5}{E_{0.95}(0.95)^2}\right)} \cdot \left(\frac{1}{0.95} - \frac{1}{0.8}\right)$$

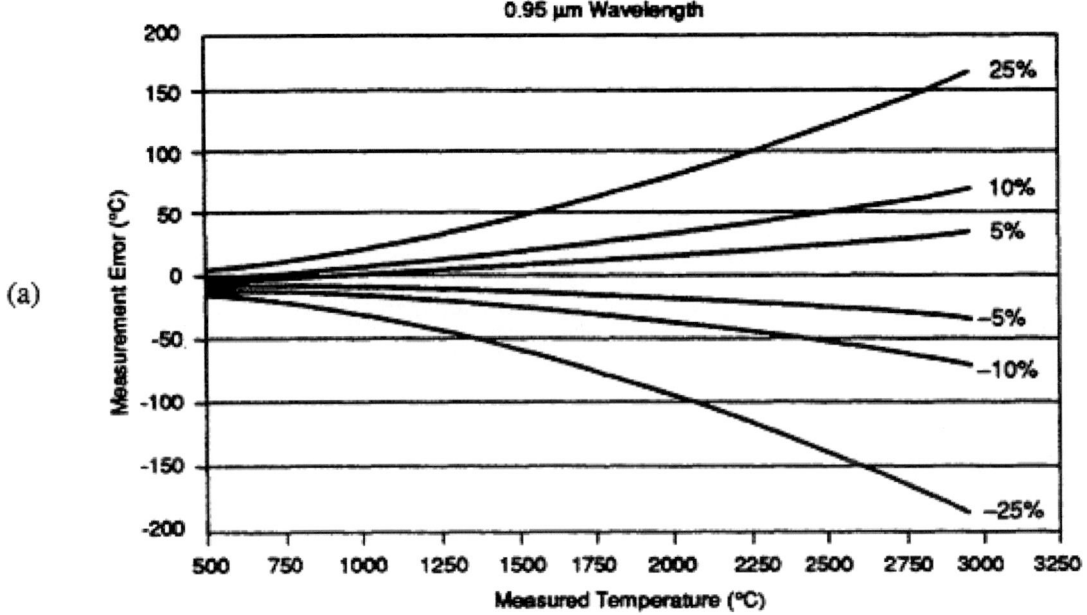

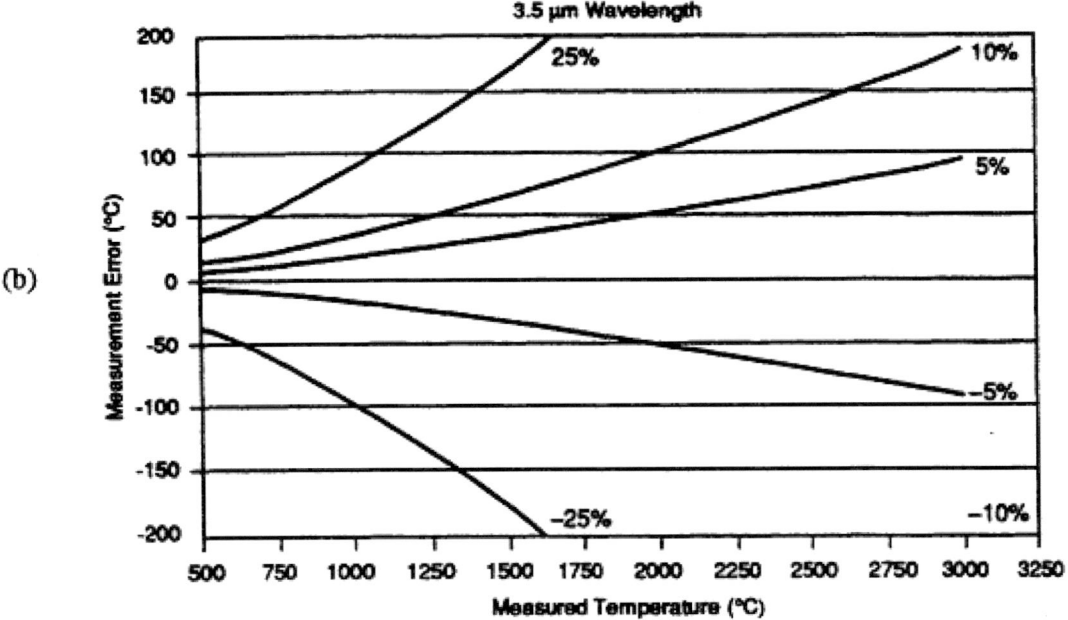

FIGURE 3-10
Measurement errors due to emissivity change with temperature and λ

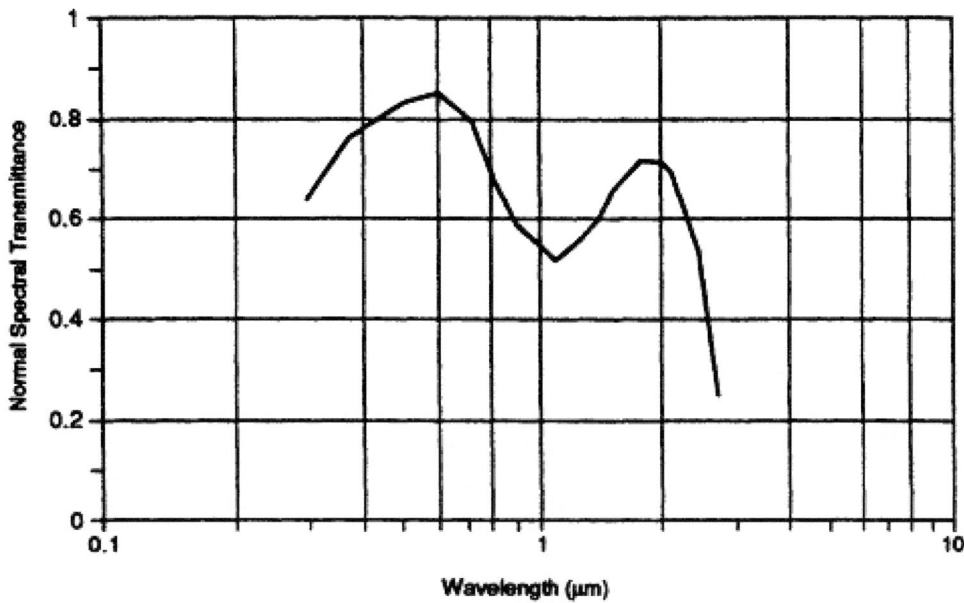

FIGURE 3-11
Normal spectral transmittance of 12.7 mm Pyrex window.

Figure 3-12 shows the dual-wavelength temperature-measurement errors caused by 10- and 25- percent difference in signal strength between two wavelengths. It may also be noted from the Pyrex window transmission graph of Figure 3-9 that a 25 percent error in transmittance is a realistic value, since transmission is 0.69 at 0.8 μm and 0.57 at 0.95 μm. From Figure 3-12 it may be seen that at a temperature of 1500 °C, a measurement error of ±270 °C may easily be encountered without even taking into account errors due to uncertainty in emissivity values.

In the case of a blackbody optical sensor, an optical fiber with a thin metallic coating at the tip, the electromagnetic interference problems that may result from the metallization at the tip of the probe, as well as the routine problem of maintaining good thermal contact between the probe and the sample, are the dominant sources of error.

Summary

Based on a brief review of available temperature measurement techniques and the complications that may result when measuring temperature in a microwave environment, the committee made the following observations:

- Temperature measurement in a microwave environment is a nontrivial procedure. Maintaining good thermal contact with the object being heated is crucial when hinting using microwaves, and it is important that temperature probes produce minimum perturbation to the existing fields in the microwave hinting chamber. This is particularly true when heating in single-mode cavities.

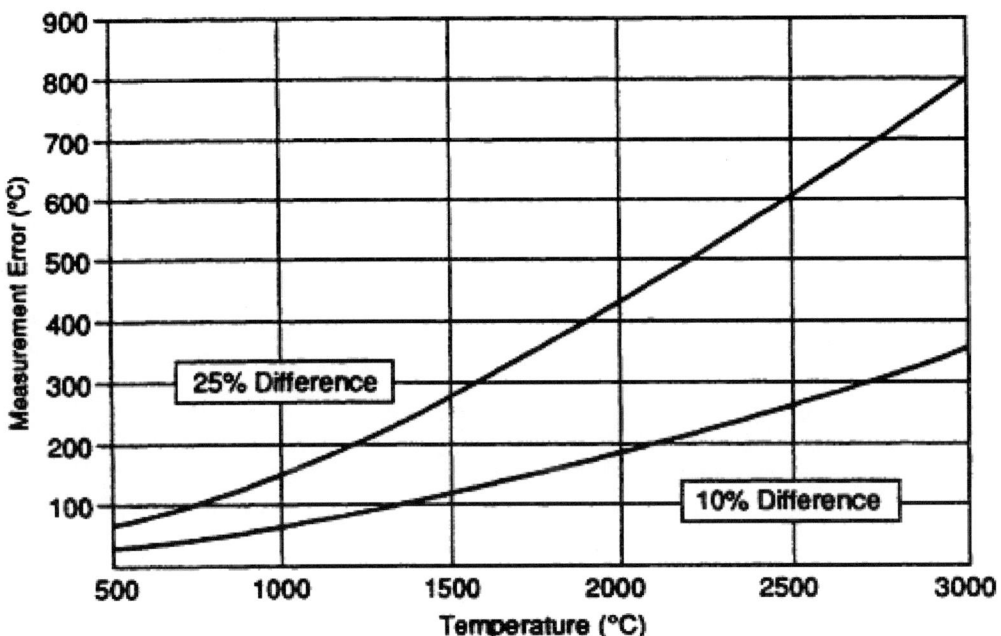

FIGURE 3-12
Temperature measurement errors for dual-wavelength optical pyrometer based on 10 and 25 percent difference in signal strength.

- Problems with probe self-heating, perturbation of fields, changing the resonant frequency in the heating chamber, window fogging, and change of transmission with frequency must be quantified and included as part of the reporting of temperature measurements.
- Procedures for estimating the effective value of the emissivity of the heated object and its variation with temperature and wavelength are important for verifying the accuracy of temperature-measurement results. These procedures should therefore be clearly described when reporting temperature measurements.
- When optical pyrometers are used in surface-temperature measurements, the transmission properties of the viewing windows and their variation with the wavelength and during the heating process must also be quantified, and their impact on the accuracy of the temperature measurement must be reported.

The committee suggests that the scientific community set a higher standard and be more critical in accepting reports of temperature measurements. The validation of reported results must be required, and a detailed description of the level of interference with the microwave environment, procedures for estimating effective values of the emissivity, the effect of sample emittance versus temperature on the estimated value of emissivity, and observation-window-related issues must all be clearly described and carefully documented.

COMPUTER MODELING AND COMPUTER SIMULATION

Computer modeling and numerical simulation can provide valuable information involving various aspects of microwave processing, including:

- microwave interaction with materials;
- effects of hybrid heating;
- heating multiple samples;
- simulating and evaluating the performance of various microwave processing systems; and
- expected magnitude and location of thermal gradients.

Valuable information regarding the uniformity of heating, the design and scale-up of a heating system, and process control parameters may be obtained from simulation results.

In spite of the many advantages of using computer modeling and numerical simulation in addressing many of the research and development and design aspects of the microwave heating process, examination of available literature reveals that reported activities in this area are rather limited (Iskander et al., 1991; Chaussecourte et al., 1992; Lorenson and Gallerneault, 1991). Other than some reported characterization of single-and multimode cavities using analytical (Manring and Asmussen, 1991; Barmatz and Jackson, 1992) and numerical (Iskander et al., 1993) procedures, most of the activities in this area were focused on laboratory experiments or empirical modeling of specific processes. Empirically simplified models and "microwave heating diagrams," based on measurements and on the data from numerical simulations conducted during development phases, can be important in the introduction of a new microwave process into a production environment. Computer simulation and numerical modeling and their use in developing and optimizing microwave processes and equipment are discussed in this section.

Numerical methods can be used to model a wide variety of microwave processing systems. For example, the finite-difference time-domain method has been used to model microwave sintering in rectangular and cylindrical single-mode cavities (Navarro et al., 1991; Chapman et al., 1992). The model used in the simulation of sintering in single-mode cylindrical cavities is shown in Figure 3-13. Model solutions allow the determination of the variation of the cavity fields with time, the time resonant frequency of the cavity, and the steady-state field and power distributions in the cavity and within the sample. Figure 3-14 shows a cross-sectional view of a calculated E-field distribution inside the cavity (with a ceramic sample inserted) and the feed waveguide at the resonant frequency. Calculations may also provide Q_s of the empty and loaded cavities, and the electromagnetic fields and power distributions at the true resonant frequency (Iskander et al., 1993; Chapman et al., 1992).

Finite-difference time-domain modeling of multimode cavities to simulate heating processes, including hybrid heating, in multimode cavities can provide information about field distributions in the cavity and about the effects of sample insertion, insulations, susceptors, etc.

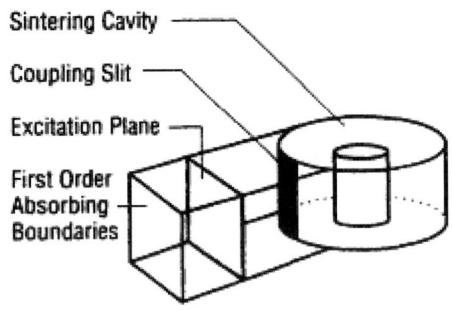

FIGURE 3-13
Model used for simulating in a single-mode cylindrical cavity (Iskander, 1993).

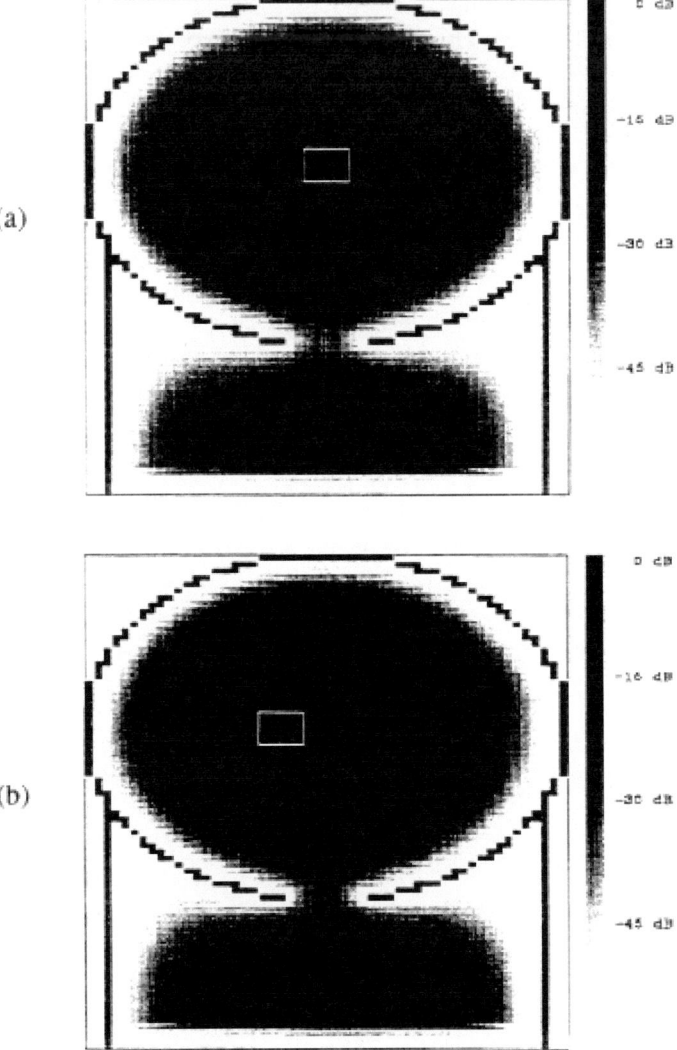

FIGURE 3-14
Cross-sectional view of the field distribution in the cylindrical cavity and the waveguide feed: (a) sample placed at the center; (b) sample placed 0.9 cm off the center.

An example of a finite-difference time-domain model traveling-wave heating system is shown in Figure 3-15 (Metaxas and Meredith, 1983). The simulation places the sample on the conveyor belt translating through the center (peak) of a TE_{10} mode in a rectangular waveguide. It is often desirable to identify guidelines regarding suitable dimensions of samples, speed of conveyor belt, input power levels, etc., using available simulation capabilities. Results of the electric field distribution in the empty applicator are shown in Figure 3-16. Figure 3-17 shows the field distribution with a sample on the conveyor belt. The standing-wave pattern (distance between peaks) is significantly different from that in the empty waveguide.

Numerical simulations can also be used to predict sample temperature and to map temperature gradients. For example, temperature profiles within a spherical sample heated by microwaves have been modeled using a shell model (Barmatz and Jackson, 1992). Field strength and absorbed power were calculated using a total-absorption model (Jackson and Barmatz, 1991). Equilibrium temperature distributions were calculated by partitioning the sphere into shells. An example of model results for alumina is shown in Figure 3-18. The model was able to predict thermal runaway behavior observed for this material (as discussed in Chapter 2).

With the availability of high-performance computing systems, graphics tools, and suitable numerical techniques, computer modeling and numerical techniques can play an important role in designing realistic microwave-heating processes. However, there are several challenges that need to be met to improve the simulation results and increase the impact on microwave processing technology.

- Results from numerical simulations need to be validated to show correlation with actual processes, including mapping of electric field and power-density distributions in microwave processing systems. As described earlier, one of the most important challenges facing the validation procedures is related to the accuracy of temperature measurements in a microwave environment.
- Accurate knowledge of the thermophysical properties of samples and insulations, including the complex permittivity versus temperature and heat transfer characteristics, is crucial to numerical simulations. The currently available data are inadequate in many cases to perform accurate simulations.
- Since large computational codes are expensive and require some expertise in computational electromagnetics, empirically simplified models and "microwave heating diagrams" should be developed based on measurements and on the extensive data collected from results of numerical simulation to make numerical techniques more accessible to processors.

Dielectric Properties Measurements

Dielectric properties measurement is an important component of the thermophysical characterization of materials. Physical interaction mechanisms between electromagnetic fields and materials can be inferred from the characteristic behavior of the complex permittivity of materials as a function of frequency and temperature. Knowledge of the dielectric properties

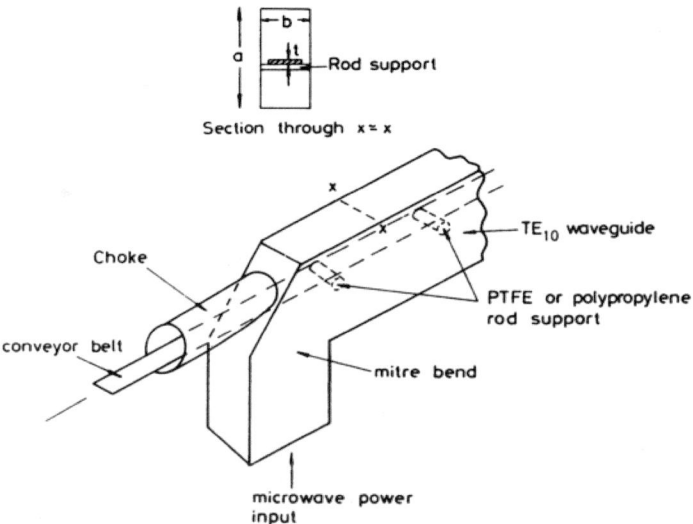

FIGURE 3-15
Model used for simulating continuous microwave heating in a traveling-wave-type applicator (Metaxas and Meredith, 1983).

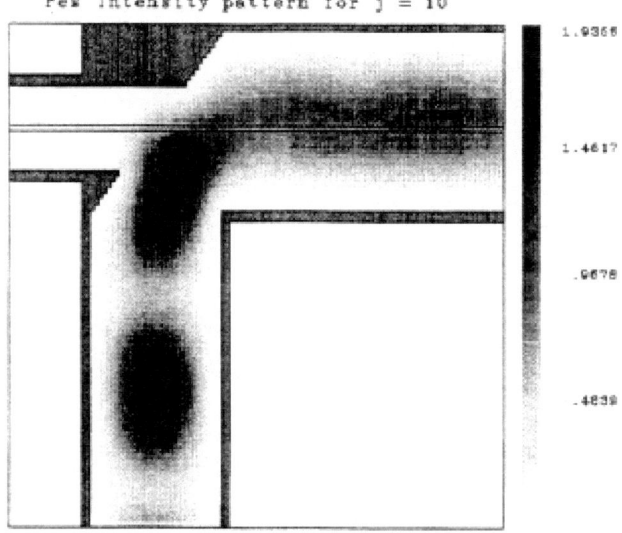

FIGURE 3-16
Electric field distribution in the empty traveling-wave applicator. The calculated standing-wave ratio as a result of the miter bend is 1.6. The standing-wave palm in the feed waveguide is illustrated.

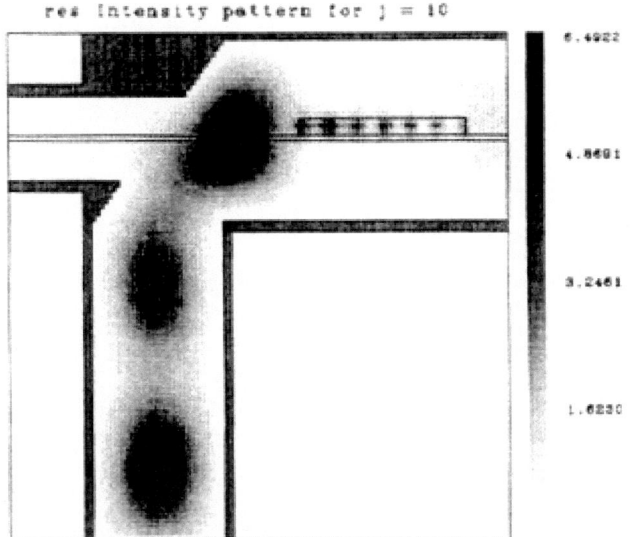

FIGURE 3-17
Electric field distribution in the traveling-wave applicator when a sample 10 cm long and 1 cm thick with $\epsilon' = 30$, σ = 0.1 S/m. The calculated standing-wave ratio of the feed waveguide was 2.3.

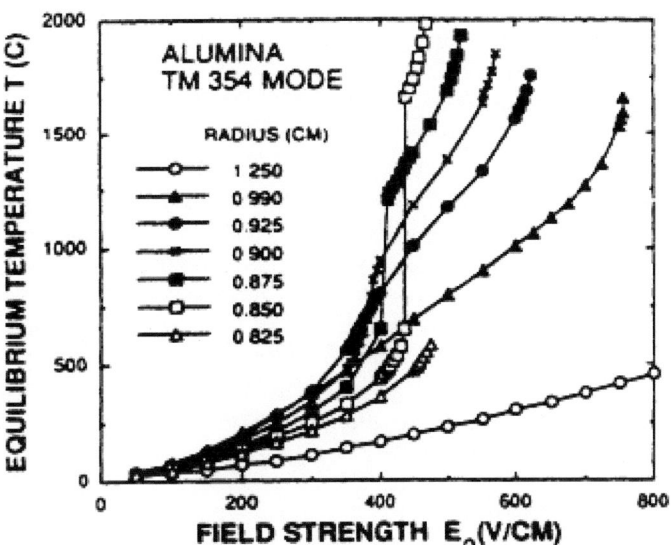

FIGURE 3-18
Equilibrium temperature at center of alumina sphere (Barmatz and Jackson, 1992).

is also important in supporting numerical modeling and calculation of the absorbed power distribution pattern in samples and insulations heated using electromagnetic energy. Even with the availability of advanced simulation and modeling techniques and software that are discussed in the previous section, application of numerical simulations is hampered by incomplete or unreliable characterization of dielectric properties of materials and their variation of dielectric properties with temperature and frequency.

Measurement of complex permittivity over a broad frequency band is required to completely characterize the dielectric properties of materials and to identify and characterize the various relaxation processes. Since several review articles on dielectric property measurement are available (Von Hippel, 1954; Westphal and Sils, 1972; Westphal, 1975, 1977, 1980; Ho, 1988), a detailed description of test methods and evaluation of their accuracy and frequency limitations is not included in this report. Rather, broadband and high-temperature measurement techniques that have been used in conjunction with microwave processing of materials——specifically transmission line, resonant cavity, and free-space methods——will be described. Transmission-line methods are the most common in the microwave band, with coaxial-line methods used in the frequency range from 50 MHz to 10 GHz and waveguide methods used from 10 GHz to 100 GHz. Resonant cavity perturbation methods provide highly accurate measurements, especially for low-loss materials, but are single-frequency measurement techniques. Free-space quasi-optical techniques are usually the most accurate for measurements above 40 GHz. Automated, vector network analyzers have enhanced electromagnetic property testing though simplification, standardization and automation of testing techniques. A network analyzer serves as a generator of electromagnetic energy and a detector of the magnitude and relative phase of the incident, reflected, and transmitted waves (Blackham, 1992).

Transmission-Line Methods

The most popular measurement techniques for characterization of complex permittivity are transmission-line methods. They provide broad band data from a single test, are relatively simple to perform, and do not require large sample sizes. In transmission-line methods, complex scattering parameters (S-parameters) of a precisely machined sample placed in a coaxial line or rectangular waveguide are determined (Weil, 1992). Use of an automated network analyzer simplifies the measurements and the determination of S-parameters. Transmission-line methods are useful in that they show relaxation behavior and transitions in permittivity with changes in frequency for the material being tested. However, there are disadvantages in using these methods. First, precise sample fit is critical, with air gaps causing significant errors. Materials that are brittle or difficult to machine are especially troublesome. Second, elevated temperature measurements using transmission-line methods are very difficult (Tinga, 1992). The entire section of transmission line containing the test specimen must be contained in an oven or furnace. Differences in thermal expansion between the sample holder and the sample under test make it difficult to maintain precise sample fit as the temperature is increased. Also, changes in the electrical properties and dimensions of the transmission lines with increasing temperature must be accounted for to maintain reference calibration (Batt et al., 1992).

The open-ended coaxial-line method for the measurement of complex permittivity has been analyzed in detail by many authors for more than a decade (Stuchly and Stuchly, 1980; Burdette et al., 1980; Athey et al., 1982; Stuchly et al., 1982; Kraszweski et al., 1983; Kraszewski and Stuchly, 1983). The open-ended coaxial probe consists of a truncated section of a coaxial line and an optional extension of a ground plane to improve the contact with the sample under test. The input port of the sensor is connected to the measurement equipment through a coaxial cable. The parameter to be measured is the admittance (or the reflection coefficient) at the interface between sensor and sample. The use of an automatic network analyzer as the measuring instrument significantly simplifies and enhances the accuracy of the measurement procedure.

The design of an open-ended coaxial probe suitable for high-temperature measurement requires the selection of a probe material that has a very low thermal expansion coefficient. A Kovar coaxial-line probe (suitable up to 600 °C), and a new probe made of metallized (silver) alumina have been utilized (Iskander and DuBow, 1983; Bringhurst et al., 1993). In the low-temperature-probe case, the calibration procedure involves the use of short and open circuits and of deionized water as standards. A modified calibration procedure, which uses short and open circuits and an alumina sample of known properties as standards, was developed to carry out calibration at temperatures as high as 1200 °C. Otherwise, the measurement procedure and the calculation approach remain the same.

Cavity Perturbation Method

Cavity perturbation methods have been widely used to measure the complex permittivity of materials at microwave frequencies. The basic assumption of this method is that the sample must be very small compared with the cavity itself, so that a frequency shift that is small compared with the resonant frequency of the empty cavity is produced by the insertion of the sample (Altschuler, 1963). The resonant frequency and Q of the cavity are determined and compared with the empty cavity values. Cavity perturbation measurements can be highly accurate and are particularly advantageous in the determination of small loss tangents.

For purposes of measuring dielectric properties at high temperatures (up to 1200 °C), the experimental setup requires the sample to be held in an adequate sample holder, which is to be heated in a conventional furnace and transferred into the cavity for test. In a typical measurement (Bringhurst et al., 1992), a thin-walled fused silica cylindrical tube was selected as a sample holder because of fused silica's temperature resistance and small variation of its dielectric properties with temperature (including low-loss). During the measurement, the tube is quickly moved from the furnace into the cavity when the desired temperature is reached. Initially, the empty sample holder is calibrated as a function of temperature (both resonance frequency and Q) to establish the "empty cavity" parameters in the perturbation expressions. Since sample and tube cooling rates are rapid after removal from the furnace and insertion into the cavity, special insulation blocks are often placed along the tube path from the furnace to the cavity to minimize cooling effects.

Free-Space Method

Free-space permittivity measurements are useful for accurate measurements at high frequency (above 40 GHz), for elevated temperature measurements, or for composite samples that have structural variations on a relatively large scale compared with sample sizes required for other measurement techniques. For this method, a plate of the sample material is placed between a high-directivity transmitter and receiver horns. When samples with relatively large dimensions are used, collimated, lens-corrected conical horns may be used to generate a near plane-wave beam over the area of interest. When sample sizes are restricted to less than the required dimensions or when spatial resolution is desired to test for sample homogeneity, spot-focusing lens antennas may be used to focus the microwave signal into a small spot on the sample. A schematic of the measurement apparatus is shown in Figure 3-19. Samples are positioned at a given angle in the path of the incident beam, and the transmission and reflection coefficients are measured by two identical receiver horns suitably aligned with respect to the incident beam and the sample. The dielectric properties are then determined from the observed transmission and reflection coefficients (Ho, 1988). Alternatively, if an automatic network analyzer is used in these measurements, either the complex transmission (magnitude and phase) or reflection coefficient may be used to determine the complex permittivity of the material under test. Measurement of the transmission coefficient is often preferred, because it avoids some measurement difficulties involved in the determination of reference planes in the reflection coefficient measurements.

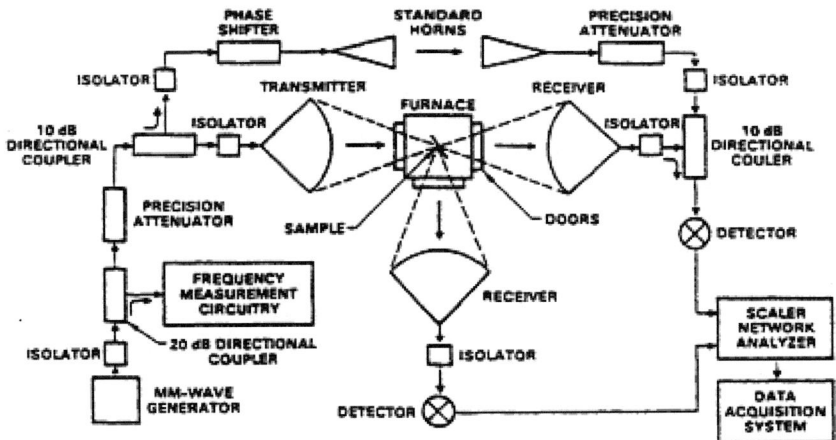

FIGURE 3-19
Free-space method for dielectric properties measurements. The transmitter, receiver, furnace, and standard horn arrangement is typical, while the rest of the measurement equipment may be replaced with a modem automatic network analysis (Ho, 1988).

4

APPLICATION CRITERIA

Microwave processing is complex and multidisciplinary in nature and involves a wide range of electromagnetic equipment design and materials variables, many of which change significantly with temperature. A high degree of technical and other (e.g., economic) knowledge is required to determine how, when, and where to use microwaves most effectively, and when not to use them (F. J. Smith, 1988, 1991; Sutton, 1993).

Commercially successful applications of microwave processing take advantage of characteristics unique to microwaves. The purpose of this chapter is to discuss those features that make microwave processing attractive for particular applications and to define the key factors that influence process economics. The goal is to provide guidance as to when microwaves can be applied to advantage in materials processing and to avoid misapplication to materials not amenable to microwave processing.

UNIQUE PERFORMANCE CHARACTERISTICS

Microwaves possess several characteristics that provide unique features that are not available in the conventional processing of materials. Some of the key characteristics of microwaves' interactions with materials are

- penetrating radiation;
- controllable electric field distributions;
- rapid heating;
- selective heating (differential absorption) of materials; and
- self-limiting reactions.

These characteristics, either singly or in combination, represent opportunities and benefits not available from conventional heating or processing methods. They also introduce problems and challenges to be met. Table 4-1 summarizes some of the features, as well as the benefits and challenges, that are associated with each of the key microwave characteristics.

APPLICATION CRITERIA

TABLE 4-1 Some Key Characteristics and Features of Microwave Processing

Characteristic	Feature	Benefits (over conventional heating)	Disadvantages
1. Penetrating radiation, direct bulk heating	• Materials heat internally • Reversed thermal gradients (ΔT) • Lower surface temperatures • Instantaneous power/temperature response • Low thermal mass • Applicator can be remote from power source	• Potential to heat large sections uniformly • ΔT favors chemical vapor infiltration; matrix infiltration • Reduced skin effect on drying • Removal of binders & gases without cracking • Improved product quality and yields • Materials & composite synthesis • Automation, precise temp. control • Rapid response to power level; pulsed power • Heat in clean environment • Materials synthesis • See differential coupling	• M/W transparent materials difficult to heat • Hot spots, cracking • Large ΔT in low thermal conductivity materials, and nonuniform heating • Controlling internal temperature • Arcing, plasmas • Require new equipment designs special reaction vessels
2. Field distributions can be controlled	• High energy concentration • Optimize power level versus time • mm-waves can be focused or defocused, rastered as desired	• Precise heating of selected regions (brazing, welding, plasma generation, fiber drawing) • Process automation, flexibility, energy saving • Synthesis of materials, composites, powders, coatings	• Equipment more costly and complex • Requires specialized equipment
3. Dielectric losses accelerate rapidly above T_{cnt}	• Very rapid heating	• Rapid processing (2-1000x factor) • Heat materials > 2000 °C • Capable to heat M/W transparent materials > T_{cnt}	• Hot spots, arcing • Nonuniform temperature • Control of thermal runaway
4. Differential coupling of materials	• Selective heating of internal or surface phases, additives or constituents	• Heating of M/W transparent, materials via additives, fugitive phases, etc. • Hybrid heating (active containers) • Materials synthesis • Selective zone heating (joining, brazing, sealing) • Controlled chemical reactions, oxidation, reduction; use of M/W transparent containers • Drying, curing, annealing; matrix infiltration	• Reactions with unwanted impurities • Contamination with insulation or other phases
5. Self-limiting	• Selective heating ceases (self regulating) after certain processes have been completed	• Below critical temperature, drying & curing are self-regulating • Completion of certain phase changes is self-regulating	• Undesired decoupling during heating in certain products • Difficult to maintain temp.

Penetrating Radiation

Microwaves can penetrate up to many meters in electrically insulating materials, such as ceramics, polymers, and certain composite materials. As discussed in Chapter 2, the depth of penetration depends on several factors, including the wavelength of the radiation and the dielectric (and magnetic) properties of the material. During the time that a material is exposed to penetrating microwave radiation, some of the energy is irreversibly lost (absorbed), which in turn generates heat within the volume (or bulk) of the material. This bulk heating raises the temperature of the materials such that the interior portions become hotter than the surface, because the surface loses heat to the cooler surroundings. This is the reverse of conventional heating, where heat from an external source is supplied to the exterior surface and diffuses toward the cooler interior regions. Thus, the reverse thermal gradients in microwave heating provide several unique benefits (Table 4-1), which include rapid volumetric heating without overheating the surface, especially in materials with low thermal conductivity; reduced surface degradation during the drying of wet materials; and removal of binders or gases from the interior of porous materials without cracking, or conversely, penetration of reactive gases (during chemical vapor infiltration) or fluids into the hotter interior portions of porous materials and preforms, then condensation into solid matter prior to the filling of voids or pores at the outer regions. These processes are discussed in more detail in Chapter 5.

Since heating is instantaneous with power input, the temperature of a material can be precisely controlled by controlling the power input. However, this is not a simple or straightforward situation, since the internal generation and surface dissipation of heat depends on many factors, which also have to be taken into account during the full heating and cooling schedule. Consequently, a detailed understanding of the microwave/material interactions, and the ability to numerically simulate, model, and predict the heating patterns for a specific material in a given microwave applicator, will play an increasingly important role in developing practical and effective controls over the various parameters for precise heating.

Since microwaves can be transmitted through air, various gases, or vacuum for long distances without significant loss in electric field strength, the power source can be remote from the applicator and sample. This makes it possible to heat in a very clean or controlled environment.

With these benefits, new problems have been encountered, such as the difficulty of trying to heat microwave-transparent materials from room temperature without generating hot spots, cracking, and arcing. At higher temperatures, the dielectric losses (and microwave absorption) of many of the materials accelerate rapidly with increasing temperature, which exacerbates nonuniform heating, warpage, and rupture problems. Many of these problems have been solved on a laboratory scale (Sutton, 1992), but much remains to be done in scaling up to an industrial process in terms of the quality, size, and dimensional complexity of the product.

Field Distribution

In a single-mode applicator (Chapter 3), the electric field distributions can be focused and controlled to provide very high field strengths. This provides a means (with proper tuning) to heat low-loss materials at low temperatures (Tian, 1991); to heat materials of irregular shape; to heat selected regions between two materials to promote welding, brazing, or bonding (Palaith et al., 1988); or to generate plasmas for microwave-assisted sintering or chemical vapor deposition (Johnson, 1992; Hollinger et al., 1993). While the single-mode processing systems allow for precise and localized heating, they are much more costly than the multimode systems and at present are not set up for large-volume processing (other than plasma applications).

At very high microwave frequencies (i.e., above 30 GHz), the wavelengths are sufficiently short that they can be beamed, focused, and rastered with metallic mirrors over desired locations on a given sample (Skylarevich and Decker, 1991). The microwave beams are far more penetrating to some classes of materials than those of lasers or electrons, so the depth of processing can be greatly increased.

Rapid Heating

For many materials, dielectric losses above a critical temperature accelerate with increasing temperature, allowing very rapid (bulk) heating that can result in very significant reductions in processing time. As discussed in Chapter 2, this rapid increase in dielectric loss can lead to uneven heating and thermal runaway. If thermal runaway is controlled using hybrid heating or insulation, this phenomenon can allow the processing of low-loss ceramic materials, where both rapid heating and very high processing temperatures are desired. The problems and possibilities associated with microwave processing of ceramics are discussed in more detail in Chapter 5.

Selective Heating

The range of dielectric responses of different materials and their ability to couple with (absorb) microwaves is one of the most widely used features of microwave processing. For example, water is a strong, broad frequency-band absorber of microwaves. This characteristic is widely used in selective heating for processing and heating food and in drying or dehydrating a wide range of wood products, chemicals, and many other materials. The processing of rubber (Krieger, 1992), asphalt (R. D. Smith, 1991), and many composite materials (Springer, 1992) is dependent on the selective or widely differential heating of at least one of the constituents. Hybrid microwave heating is another example of where selective heating has been used to significant advantage (Sutton, 1992). Examples are discussed in more detail in Chapter 5.

Self-Limiting

In several cases, microwave heating will cease once the source of differential absorption, such as water, has been removed or has been altered during a phase change in the material during processing. Self-limiting absorption can also occur when two materials with different coupling characteristics, such as SiC and ZrO_2, are simultaneously irradiated with microwaves.

At room temperature, ZrO_2 is a relatively poor microwave coupler, while SiC couples strongly. At temperatures below 300—500 °C, the SiC absorbs most of the microwave energy and is rapidly heated, which in turn heats the ZrO_2. However, above a critical temperature (approximately 500 °C), the dielectric loss factor of ZrO_2 rapidly increases with rising temperature and exceeds that of the SiC, so that the ZrO_2 absorbs increasing amounts of the microwave energy. As a consequence, the further heating of the SiC is greatly diminished (Sutton, 1992). This principle is used in hybrid heating, where susceptors are used initially to hybrid-heat low-loss materials from room temperature (Janney et al., 1991a).

ECONOMICS OF MICROWAVE PROCESSING

The usefulness of microwave energy in processing materials and the effects on material properties and yields are discussed in Chapter 5 of this report. Commercialization of the technology will be based not only on the areas of use and the properties of the materials produced but also on its economics. The need to develop ways of economically processing useful materials makes investigation of new processes valuable but means that the cost of such processes must be considered at the same time as the technical aspects.

This section of the report addresses the central question: "What are the key considerations in determining the costs associated with microwave processing of materials, and how do these costs compare with the costs of conventional processing?" An additional question that the overall report seeks to answer is: "When is microwave processing most useful or appropriate?" Answering this question requires that the process economics be addressed.

The economic feasibility is a function of local variations in energy costs, environmental laws, and labor costs balanced with the properties of finished materials or parts, improvements in yield or productivity, and the markets for the products. This report therefore seeks only to discuss the various cost aspects, provide guidelines for what must be considered, provide costs or savings as appropriate or available, and give examples as available. A definitive accounting of the costs cannot be given, because industrial microwave technology is still in its infancy, the range of application is broad, and the costs—benefit ratio will always be product specific.

It should be noted that microwave processing is unlikely to be economically competitive with processing using natural gas in the foreseeable future because of the difference in costs between natural gas (approximately $6.50 per MBTU) and electric power (approximately $17.50 per MBTU). The values cited are typical energy costs for industrial applications (Busch, 1994). Actual energy costs vary regionally.

The intrinsic performance characteristics of microwave heating have been discussed earlier in this chapter. Krieger (1989) has suggested characteristics of processes that may potentially make them attractive for microwave processing.

- The size or thickness of the material should be large.
- The cost of the material should be high.
- Improvements in properties obtainable from microwave processing are significant.
- Plant space is limited.
- Electricity is cheap.

- Minimizing handling is advantageous.

Other characteristics may include:

- heat from the combustion of coal, oil, or natural gas are not practical (ie., electricity is the only power source).
- maintaining a very clean, controlled processing environment is important.

The cost issues examined in this section include:

- cost of capital equipment, including comparison with conventional equipment;
- operating costs (energy, replacement, maintenance);
- energy required per part (energy efficiency) and cost of energy;
- deletions and addition of steps from conventional processing schemes;
- savings in time and space and changes in yield over conventional processing.

Cost of Capital Equipment

As discussed in Chapter 3, the equipment used for industrial microwave processing is generally custom designed and optimized based on specific application needs. The cost of microwave equipment depends on size, power rating, frequency, applicator design, gas control system, peripherals, manufacturer, and the size of the market for that particular equipment. Because of these dependencies, capital costs vary widely depending on the applications. Typical cost ranges are given in Table 4-2 (Sheppard, 1988). Due to the differences in the configuration and processing approach between microwave and conventional systems, it is very difficult to perform a general comparison of capital costs in a meaningful way. However, microwave processing equipment is almost always more expensive than conventional systems.

TABLE 4-2 Capital Cost of Industrial Microwave Equipment

Component	Typical Cost
Complete System	$1,000—5,000/kW
Generator	< 50% of system cost
Applicator	> 50% of system cost
Power Transmission	($1,000—3,000), $\leq$ 5% of system cost
Instrumentation	($1,000—3,000), $\leq$ 5% of system cost
External Materials Handling	($1,000—3,000), $\leq$ 5% of system cost
Installation and Start Up	5—15 % of system cost

Source: Sheppard, 1988.

Other data on microwave generator costs are given in Chapter 2 (Table 2-1).

Operating Costs

Operating costs include the cost of energy, both absolute and the real cost per part based on the coupling efficiency and the size and number of parts, and the cost of maintenance, repair, and replacement. Table 4-3 gives some estimates of these costs.

TABLE 4-3 Operating Costs of Microwave Equipment

Component	Typical Cost
Magnetron Replacement	1—12 ¢/kW·h
Electric Energy	5—12 ¢/kW·h
Plug-to-Product Efficiency	
915 MHz	70—75%
2,450 MHz	50—65%
Routine Maintenance	5—10%

Source: Sheppard, 1988.

Energy Efficiency

In analyzing energy costs for microwave processing, it is important to consider how energy is used in such operations. The theory of microwave generation and materials interactions are discussed in Chapter 2 of this report. A simplistic view is presented here only for illustration. Input or forward power is that generated by the magnetron. However, the power absorbed by the component depends on the coupling characteristics of the component, the number and arrangement of components in the cavity, and the cavity design. A certain amount of power is reflected. The absorbed energy is the difference between the input and the reflected power. It is clear, therefore, that improvements in coupling efficiency of the load material and the arrangement of insulation and components have significant effects on the energy efficiency. In typical commercial applications microwave processes have overall efficiencies of 50 to 70 percent (Metaxas and Meredith, 1983).

Energy savings arising from the use of microwave energy should be considered on the basis of the energy cost for the entire process. As discussed elsewhere in this report, hybrid systems that combine conventional and microwave heating are beneficial for many applications. Drying, for example, is best approached by first removing the bulk of the water by conventional means and then removing the remainder rapidly using microwave heating.

APPLICATION CRITERIA

The Canadian Ministry of Energy estimated potential energy savings in ceramics manufacturing using microwave energy, as shown in Table 4-4 (Sheppard, 1988). This table should be viewed with caution; the data are from before 1988, are based on growth in the industry and the entry of new companies, and are valid only for Ontario, Canada (i.e., include the efficiency of hydro-electric generation). However, the table is useful for the qualitative comparison of the amounts of energy required for different ceramic processes.

TABLE 4-4 Comparison of Energy Usage—Conventional versus Microwave Processing of Ceramics Energy usage (x106 kW.h/yr)

Product	Conventional Drying	Microwave Drying	Conventional Firing	Microwave Firing	Total Energy Savings
Brick and tile	56.10	28.05	198.90	19.90	207.06
Electrical Porcelain	3.52	1.76	12.48	1.25	12.99
Glazes	16.63	8.30	58.97	5.89	61.37
Pottery	1.96	0.98	6.94	0.69	7.23
Refractories	10.87	5.40	38.53	3.85	40.08
Sanitary Ware	25.04	12.52	88.76	8.88	92.40
Advanced Ceramics	1.30	0.65	4.60	0.46	4.79
Total	115.42	57.66	409.18	40.92	425.92

Source: Sheppard, 1988.

According to these estimates, the use of microwave drying and firing could save as much as 80 percent of the energy used in conventional processes. This figure correlates reasonably well with calculation of energy savings made in a study of sintering of alumina (Patterson, et al., 1991). The authors estimated an energy savings in excess of 90 percent when 100 components of alumina were sintered at one time.

The effects of scale-up on energy consumption for sintering of alumina are illustrated in Table 4-5 and Figure 4-1.

TABLE 4-5 Effect of Scale-up on Energy Requirements in Microwave Sintering of Alumina

Number of Components Sintered (10 g each)	Average Power (W)	Time (min)	Net energy (kJ)	Energy per component (kJ)
1	600	60	787	787
3	640	60	1919	640
5	560	60	1350	270
40	700	150	5400	135

Source: Patterson et al., 1991.

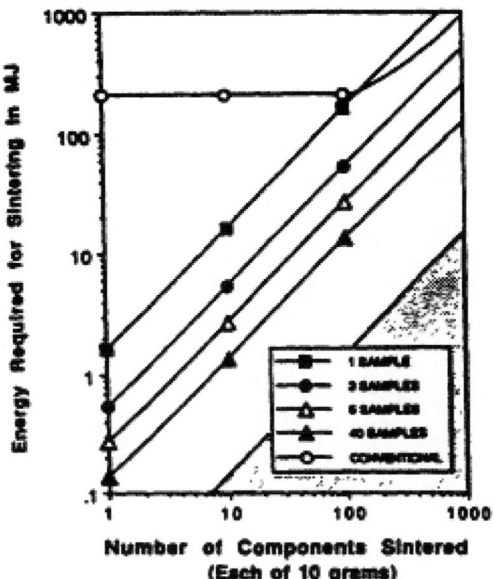

FIGURE 4-1
Energy required to sinter alumina (A16) as a function of load size (Patterson et al., 1991).

The data in Table 4-5 are for the net power, which is the input minus the reflected power, and thus may understate the amount of energy used. However, the data are useful for comparison purposes. For example, the results of this study can be compared with similar experiments at Los Alamos National Laboratory (Katz and Blake, 1991). Patterson found that the energy required to sinter a kilogram of alumina was 3.8 kW·h (in a load of 400 g). Katz found that 4.8 kW·h was needed to sinter a kilogram of alumina (250 g load). Some data on reported energy savings for microwave processing of various materials are listed in Table 4-6.

TABLE 4-6 Energy Savings Reported in Processing of Various Ceramics

Material	Process	Energy Saving Compared with Conventional Process	Performing Organization	Source
Alumina	Sintering	90 %	Alcan	Patterson et al., 1991
Ceramics	Drying	50 %	Canadian Ministry of Energy	Sheppard, 1988
Ceramics	Firing	90 %	Canadian Ministry of Energy	Sheppard, 1988
Steel Ladle Refractories	Drying	80 %	Nippon Steel	Sutton, 1992
Boron Carbide	Sintering	18%	Los Alamos	Katz et al., 1988
PZT	Sintering	95 %	Honeywell Ceramic Center	Sheppard, 1988
High Alumina Castables	Drying	20—30%	Special Metals Corp.	Sheppard, 1988

Savings from Processing Changes

Although the energy savings quoted in the previous section appear to be substantial, it should be remembered that energy costs are only a small part of the total cost of an advanced ceramic component. The greatest potential for microwave processing is in increased productivity and a consequent decrease in labor, rejection, and space costs. If improvements in properties over conventionally processed materials are realized, the premium in price obtainable for such improvements should not be forgotten.

There have been a number of reports of savings in time and improvements in productivity obtained by microwave processing (Krieger, 1989; Sheppard, 1988; Katz et al., 1988; Rains, 1988; Simonian, 1979). These are summarized in Table 4-7.

Additional savings often quoted are reductions in plant space, amount of equipment, and inventory, as well as savings in labor. The savings examples cited in Table 4-7 are for specific processes, and any estimate must be made based on a particular process.

TABLE 4-7 Time-Savings/Productivity Improvements

Material	Process	Time savings	Productivity Improvement	Performing Organization	Source
High Alumina Castables	Drying	50%	increased yield/ improved properties	Special Metals Corp	Sheppard, 1988
Whiteware	Slip Casting	66% (60 min to 20 min)	immediate mold recycling	MBM ceramics	Sheppard, 1988
Whiteware	Drying	24 h to 8 min		MBM ceramics	Sheppard, 1988
Whiteware	Overall Process	70% (7 days to 2 days)	6.25 pieces/day/worker to 27 pieces/day/worker	MBM ceramics	Sheppard, 1988
Ceramics	Drying (hybrid)	97 %		Industrial Ceramics, Ltd	Sheppard, 1988
Boron Carbide	Sintering	> 90 %		Los Alamos	Katz et al., 1988
Structural Adhesives	Curing	66%	66% cost reduction	Mobay	Rains, 1988
Varnish	Curing	<70%		Schenectady Chemicals	Simonian, 1979

Summary

The economic benefits of microwave processing are difficult to define in a general way. The decision to use microwave processing for any application has to be based on an analysis of the specific process. Important factors include the location of the processing facility; the product requirements; possible property improvements; alternative sources of energy; availability of capital; and the balance between energy costs, labor costs, capital costs, and the value added to the product.

The use of microwave processing is inhibited by the high capital costs of microwave systems and the inherent inefficiency of electric power. In most successful industrial uses of microwaves, factors other than energy account for savings realized from microwave processing; improvements in productivity and material properties, and savings in time, space, and capital equipment, are probably the best bases for selecting microwaves over conventional processes. In many applications, hybrid systems provide more savings than either microwave or conventional systems on their own.

5
MICROWAVE APPLICATIONS

INTRODUCTION

Due in large part to the overwhelming success of microwave ovens for home use, microwave processing is seen by the unwary as a panacea for all heating applications. Microwave energy is perceived to provide a means for rapid, even heating, improved processing efficiencies, and heretofore unobtainable materials properties. However, as previous sections of this report have shown, not all materials and processes are amenable to microwave processing. Even for materials and processes where microwave heating is technically an option, additional technical and economic considerations must be evaluated, on a case-by-case basis, to determine whether it is the best alternative. This chapter provides examples of work accomplished in applications of microwaves in materials processing. The observations made in previous chapters on equipment selection, process design and evaluation, and application criteria will be amplified through the examples given.

Microwave energy has found general, commercial application in very few areas. These include food processing, analytical chemistry, and heating and vulcanization of rubber. Food processing and rubber manufacture involve relatively high-volume, continuous processing. Analytical chemistry applications are broad in scope and involve high-volume, repetitive, batch processing, often with long intermediate drying and reaction steps that can be shortened using microwave heating.

Much work has been undertaken to investigate the use of microwaves for the processing of a wide range of materials, including ceramics, polymers, composites (ceramic and polymer matrix), powders, and minerals. Microwaves have also been investigated in a broad range of plasma processes (surface modification, chemical vapor infiltration, powder processing), chemical synthesis and processing, and waste remediation. Despite the considerable effort that has been expended in microwave process development, there has been little industrial application to date, with most of the effort still in the laboratory stage. Some of the more significant problems that have inhibited industrial application of microwave processing include:

- the cost of equipment;
- limited applicability;
- variation in dielectric properties with temperature; and
- the inherent inefficiency of electric power.

Much of this work has been undertaken without the initial cross-disciplinary evaluation and processing system design approach emphasized in this report. Discussion of these results in light of this multidisciplinary approach will serve to highlight the limitations in terms of capabilities and scaling and will lead to identification of promising processes and needed research.

A broad range of applications will be discussed. Much work has been accomplished on ceramic, polymer, and plasma processing, and the lessons that can be learned from this work will help to identify promising applications for future development and will help processors avoid possible pitfalls. Emerging and innovative applications in microwave chemistry, minerals processing, and waste remediation are also reviewed.

CERAMICS/CERAMIC MATRIX COMPOSITES

The use of microwave energy for processing ceramics and ceramic matrix composites has been the subject of a large amount of exploratory research. The range of materials and processes that have been investigated is shown in Table 5-1.

TABLE 5-1 Examples of Ceramic Microwave Processing Research and Development

Materials	Annealing	Binder Burnout	Calcining	Drying	Fiber Drawing	Joining	Melting	Sintering	Combustion Synthesis	Powder Synthesis	Slip Casting	Clinker
Advanced Ceramics		X		X		X		X	X	X		
Cements												X
Composites						X		X				
Ferroelectrics	X	X	X					X				
Ferrites								X				
Glasses					X		X			X		
Minerals			X	X				X				
Refractories			X	X				X	X			
Superconductors	X			X				X	X			
Whitewares				X				X			X	

The potential advantages of microwave processes over conventional processes for ceramic processing include reduced processing time, improved product uniformity and yields, improved or unique microstructure, and the ability to synthesize new materials (Sutton, 1989). A number of review articles on ceramic microwave applications have been published (Sutton, 1989, 1993). Rather than providing a broad review of ceramic processes, this section will examine two important processing areas——sintering and powder processing——in light of the perceived advantages described above, comment on lessons to be learned from previous work in these areas, and suggest promising applications for the future. Microwave sintering is of interest because of the extensive exploratory work accomplished and because of the broad range of ceramic materials that have been investigated. Microwave processing of ceramic powders is a relatively new area with promise of broad applications in synthesis and processing.

Sintering

Microwave heating has been touted as a means of sintering ceramics since the early 1970s. Microwave sintering of a number of oxides and nonoxide ceramics ranging from low-loss materials like Al_2O_3 to relatively high-loss materials such as SiC, TiB_2, and B_4C has been reported. The perceived advantages of microwave sintering over conventional sintering include expectations for more-uniform heating, better properties of the product, greater throughput with resulting smaller plant size, and greater energy efficiency. It is generally assumed that since microwave energy is deposited in the bulk, significantly less time is required to heat the part to the sintering temperature than would be required to diffuse the heat from the exterior, particularly for large parts or large batches of small parts. The resulting rapid sintering may lead to smaller grain size at a given density, with consequently better mechanical properties. Although some advantageous application of microwave processing in sintering has been demonstrated, the perceived potential of the technology has gone largely unrealized on a production scale.

Oxides

Berteaud and Badot (1976) investigated the sintering of alumina and zirconia and the melting of silica at 2.45 GHz in a rectangular single-mode cavity. They recognized many of the potential advantages of microwave sintering, including high thermal efficiency as well as rapid processing, and also discovered many of the problems that have plagued the process, including difficulty in temperature measurement due to temperature gradients and the propensity for thermal runaway. Colomban and Badot (1978, 1979) investigated the sintering of β-alumina, again in a single-mode cavity, where they observed rapid sintering but not the expected small grain size.

Microwave sintering of alumina sparkplug insulators was investigated with goals to replace large (50-m long), gas-fired line kilns with significantly smaller equipment and to reduce process cycle-time (Schubring, 1983). Microwave sintering was found to be feasible, with cycle time reduction from 24 hours to 3—6 hours. The energy consumption was half that for gas firing, but the energy costs were higher for microwave heating because of the relative costs of gas versus electric energy. Although part to part density variations were observed in the 186-part sagger, acceptable properties were obtained. A similar study of the sintering of ferrites resulted in similar conclusions regarding feasibility (Krage, 1981). However, neither process was carried to practice.

Large castable refractory crucibles have been successfully sintered in a microwave cavity (Sutton, 1988). Firing times were significantly reduced compared with conventional heating, because the penetration depth of microwaves allowed even heating throughout the thickness of the rather large sections involved.

Ultra-rapid sintering of β-alumina, Al_2O_3, and Al_2O_3/TiC rods and thin-walled tubes using a rapid pass-through, zone-sintering process in single-mode applicators, has been investigated (Johnson, 1991). Isostatically pressed rods, 4 mm in diameter, of β-alumina powder were sintered at specimen translation rates of up to 40 mm/min, with the final density independent of translation rate. The time from onset of heating to final density was on the order of 30 s at the highest translation rate. Attempts to sinter thin-wall β-alumina with a diameter of 15 mm tubes failed, because a small region of the tube would become hot and remain hot to the exclusion of the rest of the specimen, even though the tube was rotated in the cavity. The size of the spot, on the order of several millimeters in diameter, was sensitive to the power applied, was stable in time, and did not propagate around the circumference of the tube.

A single-mode cavity was used to sinter α-alumina rods with a diameter of 4 mm to high density and fine grain size (99.8 percent dense and 2 μm, respectively; Tian et al., 1988a). To avoid thermal runaway, applied power had to be carefully controlled as the sintering temperature was approached. Stable heating, again with high densities and fine grain size, was also observed in sintering Al_2O_3/TiC rods with a diameter of 4 mm (Tian et al., 1988b). Thermal runaway was avoided if the concentration of TiC was greater than about 20 percent by weight.

Of several reported attempts to sinter Al_2O_3 in multimode cavities, the experiments of Patterson et al. (1991) were among the most successful. Single and multiple specimens (19 mm diameter by 16 mm long) were sintered to greater than 98 percent density with three different alumina powders. If the heating rate was too high, nonuniform grain sizes resulted, with the largest grains in the center of the specimen. A 60-minute firing cycle that resulted in uniform grain sizes was developed. Few details about procedures, thermal insulation, or the oven configuration were given.

Sintering of a few other oxide materials with varying degrees of success has been reported. Some of these reports are listed in Table 5-2. In most cases, the procedures were not described well enough for the committee to determine whether sintering enhancement was observed.

Nonoxides

B_4C and TiB_2 have been successfully heated to very high temperatures using granular Y_2O_3 as the microwave-transparent primary insulation system (Holcombe and Dykes, 1991a, b). Increased density and improved mechanical properties of microwave-sintered B_4C were reported

TABLE 5-2 Selected Microwave Sintering reports.

Material	Insulation	Coupling	Reference
Al_2O_3	None	Self	(Tian et al., 1988a)
Al_2O_3	Al_2O_3 fiber	Hybrid (insulation + self)	(Patil et al., 1991)
Al_2O_3	Un-named fiber	Hybrid (SiC liner)	(Dé et al., 1991b, c)
Al_2O_3	Not given	Not given	(Patterson et al., 1991)
Al_2O_3	Un-named fiber	Hybrid (tubular receptors)	(Brandon et al., 1992)
Al_2O_3/MgO	Al_2O_3-SiO_2 fiber	Not given	(Cheng et al., 1992)
Al_2O_3-TiC	None	Self	(Tian et al., 1958b)
Al_2O_3-ZrO_4	Al_2O_3 and/or ZrO_2	Hybrid (picket fence, 2.45 GHz); self at 28 GHz	(Kimrey et al., 1991)
Al_2O_3-ZrO_4	Un-named fiber	Hybrid (tubular susceptors)	(Brandon et al., 1992)
ß-alumina	None	Self	(Johnson, 1991)
B_4C	Y_2O_3 grain	Self	(Holcombe and Dykes, 1991a)
$BaTiO_3$	Y_2O_3 fiber and powder	Probably self	(Lauf et al., 1992)
Hydroxyapatite	Zr_2O_3 fiber	Not given	(Agrawal et al., 1992)
$LaCrO_3$		Self	(Janney and Kimrey, 1992)
Si_3N_4	ZrO_2 and Safil fibers	Hybrid (powder bed)	(Patterson et al., 1992a)
Si_3N_4	Not given	Not given	(Zhang et al., 1992)
TiB_2	Y grain	Self	(Holcombe and Dykes, 1991b)
TiO_2 nanophase	ZrO_2 fiber	Not given	(Eastman et al., 1991)
$YBa_2Cu_3O_x$	Al_2O_3-SiO_2	Hybrid (SiC liner)	(Ozzi et al., 1991)
ZnO varistor	Not given	Probably self	(Levinson et al., 1992; McMahon et al., 1991)
ZrO_2/12% CeO_2	ZrO_2 fiber	Hybrid (insulation)	(Janney et al., 1992b)
ZrO_2/8% Y_2O_3	ZrO_2 fiber next to specimen	Hybrid (insulation, picket fence; self at 28 GHz	(Janney et al., 1991b, 1992b)

compared with the conventionally sintered material. A barrier layer was required to preclude Y_2O_3 contamination of the TiB_2.

Batches of Si_3N_4 cutting tools, with 90 parts per batch, were sintered using a cylindrical multimode cavity (Patterson et al., 1992a). Parts were arranged in six layers embedded and isolated from each other within a packing powder and were enclosed in a cylindrical alumina crucible. The packing powder, consisting of 40 percent SiC, 30 percent BN, and 30 percent Si_3N_4 by weight, served multiple purposes——providing a source for N_2, providing high thermal conductivity, acting as a getter for O_2, and acting as a microwave absorber. Three conductive rings were placed around the alumina crucible to shape the microwave field. The temperature increased with some degree of nonuniformity during a slow increase in microwave power. A locally high-temperature area would commence at one end of the load and gradually spread throughout the entire load. Thus, after 50 minutes the surface temperatures ranged from 536—1190 °C, whereas after 140 minutes the range was from 1540—1610 °C. After optimizing the process, uniform density among the parts was obtained. Energy consumption was estimated to be on the order of 80 percent less than experienced with conventional heating. In this case, electric heat is mandated for conventional processing because of the tendency of the material to oxidize in combustion gases.

Issues in Microwave Sintering

Microwave Enhancement Effects

There have been numerous reports of enhancement of sintering kinetics when using microwave processing. Probably the most startling is a report of as much as a 400 °C reduction in sintering temperature along with a dramatically reduced activation energy for Al_2O_3 processed in a 28-GHz microwave cavity (Janney and Kimrey, 1988, 1990). As discussed in Chapter 3 of this report, significant errors in temperature measurement can lead to misleading processing results. Shielded and grounded thermocouples, as discussed in Chapter 3, as well as optical pyrometers, were used to minimize temperature-measurement errors, and carefully designed insulation systems were used in the studies referenced above to minimize temperature gradients. By switching the microwave power off and on, and observing thermocouple response, it was demonstrated that the microwave field did not bias the thermocouple output (Janney et al., 1991a). While temperature measurement may yet be a problem, it is difficult to imagine a 400 °C error.

Significant reductions in sintering temperatures or enhancements in the diffusion coefficient for sintering have also been reported for Al_2O_3 (Patil et al., 1991; Cheng et al., 1992) and Al_2O_3 doped with MgO (Cheng et al., 1992). Reductions in sintering temperature and activation energy have been reported for the sintering of zirconia-toughened alumina (Kimrey et al., 1991) and zirconia (Janney et al., 1991b, 1992a). A variety of other ceramics were similarly sintered using microwaves, including B_4C (Holcombe and Dykes, 1991a), $LaCrO_3$ (Janney and Kimrey, 1992), and Si_3N_4 (Tiegs et al., 1991; Kiggans et al., 1991; Kiggans and Tiegs, 1992). These results, for a broad range of materials, indicated that the reduction in sintering temperature was observed in insulators and ionically conducting materials but not in electronically conducting materials. Reductions in sintering temperature and activation energy were greater at 28 GHz than they were at 2.45 GHz.

Enhanced microwave plasma sintering of alumina and a few other oxides has been observed, but only the data for alumina were presented quantitatively (Bennett et al., 1968). A 200 °C reduction in the sintering temperature of Linde A alumina was reported compared with the same temperature and time in a conventional furnace. Rapid pass-through sintering of thin-wall tubes and rods in the radio frequency (RF) induction coupled plasma and microwave plasmas has been investigated (Sweeney and Johnson, 1991). Densification times in thin-wall tubes in the RF induction coupled plasma were as low as 10 seconds from onset to completion of densification, with final densities as high as 99.7 percent for MgO-doped alumina. Similar sintering speeds were obtained with rods in a microwave plasma. Although the microwave plasma process has not yet been thoroughly characterized, a clear enhancement of sintering was observed in the 5-MHz induction coupled plasma sintering of alumina (with pains taken to correct for temperature measurement errors through extensive calibration procedures).

Microwave enhancement effects have not been observed universally. Patterson et al. (1991) saw a slight increase in the sintering rate of three alumina powders——the sintering time was cut in half at 1600 °C relative to conventional sintering, with comparable densities and elastic modulus. Levinson et al. (1992) found no significant difference in the sintering of ZnO varistor materials in microwaves relative to that in conventional firing, and there was no difference in properties. The interdiffusion of Cr_2O_3 and Al_2O_3 under microwave heating was studied to determine if there was enhanced diffusion in ceramics heated by microwaves (Katz et al., 1991). Although a slight apparent enhancement was observed, it was concluded that this could be accounted for without resorting to a rate enhancement by microwaves.

Part of the controversy surrounding the "microwave effect" is that satisfactory physical explanations are missing. Booske et al. (1992) proposed a theory in which the enhanced sintering is attributed to enhancement of the phonon energy distribution in the high end of the distribution. The same research group later reported that further calculations showed the proposed effect was of insufficient magnitude to explain the observations (Booske et al., 1993). A satisfactory physical explanation of microwave effects must show why electronically insulating materials have shown the effect while conducting materials have not.

A series of careful experiments is needed to eliminate the doubts that remain about the "microwave effect." Since temperature measurement is often problematical, some method of internal calibration of the temperature is imperative.

Hybrid Heating

Electrically transparent (low-loss) materials, such as SiO_2 and Al_2O_3, are difficult to heat at room temperature. Additionally, many materials that are hard to heat at room temperature possess electrical conductivity or dielectric loss factors that rise rapidly in magnitude as the temperature rises. Thus these materials will absorb microwave energy if they can be preheated to a suitable temperature using another source of heat. This has led to the development of passive hybrid heating using higher dielectric loss susceptors, insulation, or coatings that absorb incident microwave power more readily at low temperature.

The sintering of ZrO_2-toughened Al_2O_3, ZrO_2/8% Y_2O_3 and zirconia/12% CeO_2 at 2.45 GHz provides an example of how a hybrid heating process can improve unstable heating (Janney et al., 1992b). Although these materials could be sintered readily at 28 GHz, attempts at 2.45 GHz, where equipment costs are more attractive, were frustrated until SiC rods were inserted into the insulation that surrounded the specimens in what was referred to as the "picket fence" arrangement. The microwave energy initially heated the SiC rods, which, in turn, transferred heat to the insulation and eventually to the specimens. As the specimen temperature increased, it more effectively coupled with the microwave energy and began to heat directly. Figure 5-1, part a shows the calculated field distribution (using finite-difference time-domain modeling) in a simulation of four ceramic samples surrounded by insulation inside the cavity, showing a concentration of fields at the sides of the insulation. Although the inclusion of four SiC rods, as shown in Figure 5-1, part b, reduced the overall field strength, it helped to concentrate the fields in areas of interest near the samples rather than in the insulation. With this arrangement, these materials were successfully sintered, albeit at fairly low heating rates (2 °C/min).

Another hybrid microwave heating scheme involved applying a thin layer of SiC powder to the interior of the thermal insulation chamber that is placed within the microwave oven (Dé et al., 1991a, b, c). As in the picket fence arrangement, the silicon carbide is initially heated by the microwaves, transferring heat to the specimen. The silicon carbide layer is thin enough that significant penetration of the microwaves occurs. With this arrangement, a number of ceramic materials have been successfully sintered.

In yet another hybrid heating process, tubular susceptors of various sizes were inserted over relatively large sized compacts of alumina and zirconia toughened alumina (50 mm diameter by 60 mm long) which were then sintered to 1500 °C at heating rates of about 10 °C/min (Brandon et al., 1992). Compacts of this size could not be sintered using conventional processes at 10 or even 5 °C/min without cracking.

Simulations have shown that increasing the ambient temperature through some form of hybrid heating can increase the critical temperature to as high as required for sintering. These results explain the success achieved with hybrid heating processes that is reported in the literature. The simulations also explain, qualitatively at least, the observed difficulties in sintering low-loss oxide materials by microwave heating (Spotz, et al., 1993).

Although some impressive results have been reported in the hybrid heating of alumina, controlled rapid heating of oxides with both low initial dielectric loss factors and high temperature dependence of dielectric loss factors is difficult to achieve. Success generally has been limited to single specimens of simple geometry in carefully designed sintering chambers.

Insulation

Unstable heating due to changing permittivity or thermal gradients caused by heat loss from part surfaces can be minimized using effective insulation. Almost all cases in which successful microwave sintering has been reported have necessitated carefully designed insulation systems. In many cases, packing powders are required, which would only be acceptable for products of very high value. Development of a workable insulating system has been identified

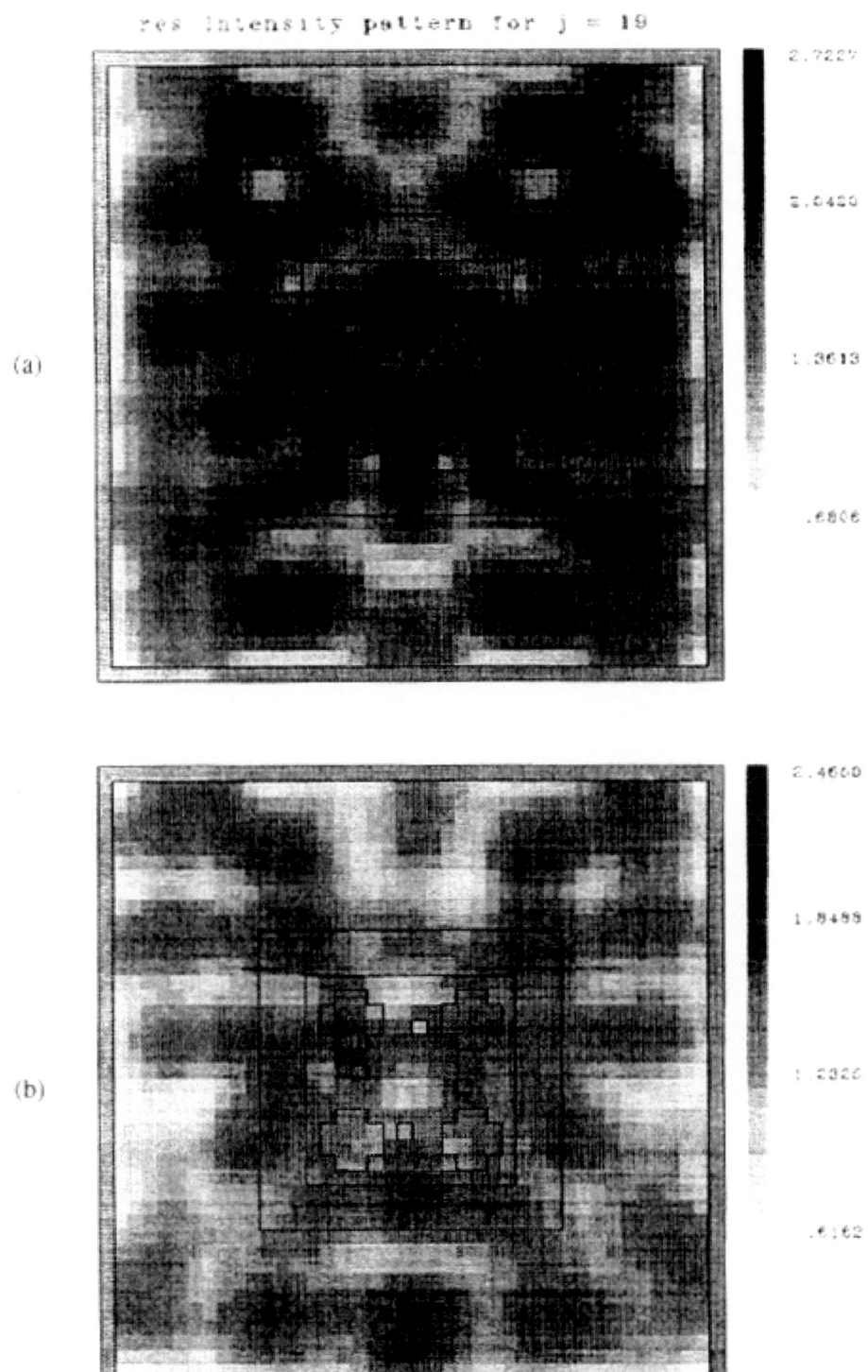

FIGURE 5-1
Calculated electric field distribution in a multimode cavity when four ceramic samples ($\epsilon' = 4.13$, $\sigma = 64 \times 10^{-6}$ S/m) are surrounded by insulation. (a) samples alone; (b) samples plus SiC rods.

as "one of the most challenging tasks in the high temperature microwave processing of ceramics" (Janney et al., 1992b).

While temperature gradients within the specimen can be reduced by the presence of insulation, they can be eliminated only if all of the microwave energy is absorbed by the insulation, or a susceptor, and subsequently transferred to the specimen. In fact, the various hybrid heating approaches move in this direction. However, temperature gradients of a certain magnitude may be acceptable.

If the microwave power absorption increases sharply with increasing temperature, there always will be heating problems because of this volumetric heating and surface cooling of even well-insulated specimens. In some cases, surface thermal gradients may be overcome by slow heating, but, in other cases, microwave heating will be unstable, making uniform sintering impossible. This difficulty is particularly exacerbated in the case of multiple-part sintering, where a hotter part may preferentially absorb more microwave energy at the expense of cooler parts.

Materials with well-behaved heating behavior, such as the carbides, usually require very high sintering temperatures. This presents problems with regard to setters and insulation. There are no truly microwave-transparent insulation materials capable of operation in the 2000 °C range, although granular Y_2O_3 has shown some promise.

The need for carefully controlled insulation has forced microwave sintering to be basically a batch process, often with only a single part being sintered at a time. Successful demonstration of large batches has only rarely been successful. Even in successful cases, part-to-part variations were common. In the extreme cases, some parts have undergone thermal runaway and others were not fully sintered. Unfortunately, these effects are exacerbated by large batch size and rapid heating, both of which are desirable from a manufacturing point of view. Further investigation is needed to discover the regimes of microwave-power absorption characteristics, batch size, heating rate, and other variables in which microwave sintering can be reproducible and uniform.

Thermal Runaway

As discussed in Chapter 2, the rapid rise in dielectric loss factor with temperature is the major issue in thermal runaway and temperature nonuniformity. Therefore, although microwave heating frequently is touted as providing more uniform heating, nonuniform heating is a reality in many oxides, often at nominal heating rates. The situation is worse when a multitude of parts are heated together, or for other than simple specimen geometry.

Some general observations can be made about factors relating to thermal runaway. First, if the temperature dependence of the power absorption is less than the temperature dependence of the heat dissipation at the surface of the specimen plus insulation system, stable heating should be observed. Second, hybrid heating using either lossy insulation or other susceptors that absorb a significant fraction of the microwave power and transmit it to the specimen by conduction or radiation is important in stable heating. Finally multiple specimens of differing size, or specimens with varying cross section or complex shapes, will be particularly difficult to heat uniformly. Materials with lower temperature dependence of dielectric loss factor may be heated stably. However, the uniformity issues for complex shapes or differing sizes within a batch will persist. Further work is required to determine more fully the conditions under which stable heating of various materials can be achieved.

Property Enhancement

The final issue is the question of whether there are fundamental differences in the properties achievable by microwave sintering and those achievable by other methods. Microwave sintered and hot isostatically pressed Si_3N_4 cutting tools showed significantly improved performance compared with commercially available cutting tools (Patterson, 1992b). Dé et al. (1991b, c) investigated the effects of heating rate on the densification and microstructure of conventionally and microwave sintered materials in a hybrid system. They observed that higher heating rates result in higher density and smaller grains, just as with conventional fast firing. However, higher heating rates were achieved in the hybrid system than was possible with the same specimen size in a conventional furnace. It may be significant that the microwave sintered specimens had a smaller grain size at any given density during the densification process than did conventionally fast-fired specimens. Unfortunately, other researchers have not reported relationships of grain size versus density that would make it possible to determine whether this effect is widely realized with other materials.

Powder Processing

The synthesis and processing of powders is a key technology area affecting the future development of advanced ceramic materials. The application of microwaves to powder processing technology is relatively new and will be discussed briefly. Table 5-3 summarizes some of the areas where microwaves have been applied to ceramic powder processing.

Powder Synthesis

The characteristics of a starting powder (composition, size, structure, shape, etc.) strongly affect the control over the sintering behavior, microstructural development, improved properties, and reliability of the final product (Johnson, 1987). For this reason, there continues to be a significant effort to develop improved and tailorable powders to meet the increasing demands for a wide range of future, advanced ceramic products (Messing et al., 1987, 1988a, b).

The application of microwaves to the synthesis of ceramic (oxide and nonoxide) powders is a recent and emerging development and offers some unique benefits, especially with respect to producing particles of submicron (nano) size with controlled compositions. Microwave synthesis of ceramic powders offers greater process flexibility by taking advantage of several combinations of volumetric, rapid, and selective heating conditions, which are not possible by conventional means. All of these heating advantages can be used to process and tailor extremely fine (less than 1-μm diameter) powders by controlled reactions in sol-gel processing, gas-phase synthesis, solution evaporation/decomposition, or hydrothermal reactions. Each of them, and other powder synthesis methods, will be described next.

TABLE 5-3 Microwave Applications in Ceramic Powder Processing

Powder Synthesis
- Sol-gel Decomposition/Drying
- Solution Evaporation/Decomposition
- Gas-Phase Reactions
- Gas-Solid Reactions
- Solid-State Reactions
- Ceramic Precursor Pyrolysis
- Hydrothermal Reactions

Powder Treatment
- Dissolution
- Drying
- Calcining

Powder Consolidation/Shaping
- Sintering
- Reaction and Sintering
- Melting
- Ignition

Sol-Gel Decomposition/Drying

Microwaves have been used in several of the processing stages to synthesize $BaTiO_3$ powders from a sol-gel precursor. A solution of barium and titanium acetate was decomposed to produce a dry gel, which was pyrolyzed to yield a brown product and then calcined to yield a colorless powder sample of $BaTiO_3$ (Kladnig and Horn, 1990). Microwave energy was effectively utilized in all of these stages.

Fine crystalline mullite powders have been prepared from single-phase gels in a few minutes (Komarneni et al., 1988). The particle sizes of the dry powders were about 0.1 to 0.5 μm, with mullite crystallite sizes of 100—200 nm after microwave heating the gel for 5 minutes. At present, the mechanism for the microwave absorption of the aluminosilicate gels is not well understood. In other sol-gel studies, microwave absorption was also significant, and silica (Roy et al., 1985) and urania (Haas, 1979) gels could be rapidly dried and heated to their melting points.

Solution Evaporation/Decomposition

This method can also produce extremely fine powders of controlled (mono-or polyphased) compositions and high purity. A novel approach to this process used microwaves to decompose aqueous solutions of nitrates, nitrate-HF, and chlorine after the solutions had been sprayed into a microwave chamber (Kladnig and Horn, 1990). Water vapor and other volatiles were removed via a vacuum to pass into an absorption (condensing) system, which would regenerate the solvents (HNO_3, HCL, HF, etc.). Depending on the composition of the starting solution, powders of ferrites, Al_2O_3, TiO_2, and other oxides were produced. One advantage of this method is that it could be developed into a continuous process with the recovery of some of the starting solvents.

Microwave-generated plasmas have also been used to decompose atomized droplets of aqueous solutions containing the nitrates of Zr and Al, yielding very fine crystalline powders of γ-Al_2O_3 and ZrO_2, respectively (Vollath et al., 1992). When nitrates of Zr, Al, and Y were atomized together, a mixed-oxide powder of ZrO_2, Al_2O_3, and Y_2O_3 was produced. The chief advantages of microwave plasma processing were high efficiency (80 percent) in transferring thermal energy to the chemical reactions; formation of completely crystalline, spherical particles of ZrO_2 (100—500 nm in diameter); and a capability to produce solid-solution particles.

Gas-Phase Reactions

Nonoxide powders of AlN, SiC, and Si_3N_4 have been synthesized by nonthermal microwave plasmas of precursor gases under conditions of laminar flow (Singh et al., 1991). The product particles were ultra-fine (~ 5 nm) and crystalline. AlN was stabilized in either the hexagonal or cubic phase, depending on the nitrogen concentration during the reactions. The SiC formed mostly cubic-3C, with other polytypic modifications, while the Si_3N_4 was formed as an α-phase modification.

Gas-Solid Reactions

This method has not received much attention from a microwave processing viewpoint. However, in a study of processing refractory ores in a microwave-induced cold plasma, Bullard and Lynch (1992) investigated the reduction of TiO_2 powder in a hydrogen plasma under reduced pressure (16 Torr). They observed about a 60 percent conversion to the Ti_2O_3 phase in 11 minutes at low temperatures (below 735 K).

Solid-State Reactions

Microwaves have also been used to promote reactions between mixtures of solid particles to form powders of new (reacted) compositions.

In the case of oxide powders, a variety of oxides, such as KVO_3, $BaWO_4$, and $YBa_2Cu_3O_{7-x}$ have been produced via solid-state microwave synthesis (Mingos and Baghurst, 1992). These authors also used microwaves in the synthesis of borides by heating mixtures of boron with Cr, Fe, and Zr to 1000 °C.

Ultra-fine SiC powders have also been synthesized by the carbothermal reduction of silica using microwave and conventional firing techniques (Kumar et al., 1991). Both techniques produced ß-SiC powders, but the crystallite size of the microwave-produced powders was 30—200 nm, versus 50—450 nm for the conventionally produced material. Microwaves were also used to synthesize SiC, TiC, NbC, and TaC from mixtures of the corresponding metal oxides and graphite powders. Temperatures of up to 1400—1500 °C were obtained in 13 min, and the carbides were formed within 20 min (Kozuka and MacKenzie, 1991). This technique also appears to be a new means to produce SiC whiskers.

Ceramic Precursor Pyrolysis

A wide variety of ceramic powders have been produced by microwave heating of ceramic-precursor compounds or mixtures of such compounds without added solvents, thus avoiding the large volume of solvents to be removed in the solvent decomposition/evaporation process used in conventional processing (Willert-Porada et al., 1992). Control of powder properties is achieved through chemical modification of reaction mixtures, use of specially designed microwave applicators, and control over certain decomposition profiles. Fine powders were produced by using ceramic precursor alcoholates and acetylacetonates of Al, Zr, Ti, Si, Cu, and Mg. These compounds absorb microwaves readily. Single oxide powders, such as Al_2O_3 and ZrO_2, were prepared by pyrolysis of Al-triisopropanolate (ATIP) or Zr-tetrapropylate (ZTP). Mixed-oxide powders, such as Al_2O_3 + ZrO_2, Al_2O_3 + CuO or $CuAlO_2$, or $MgAl_2O_4$, were prepared by the pyrolysis of appropriate precursor mixtures.

In addition, composite powders were also prepared by coating inert particles of Al_2O_3, BN, or SiC with a thin layer of a zirconia precursor (ZTP) or by coating reactive powders such as carbon with ZTP and other precursors to form carbide/oxide composite powders (Willert-Porada et al., 1992). As shown in Figure 5-2, by using microwave heating of metallorganic precursors, decomposition is enhanced and occurs at lower macroscopically measured temperatures than conventional thermal processing, so that a wide selection of mono- and polyphasic powders could be synthesized with reasonably high surface areas (10—700 m^2/g).

Hydrothermal Reactions

Microwave-hydrothermal processing has been utilized in catalyzing the synthesis of crystalline, submicron powders of unary oxides such as TiO_2, ZrO_2, and Fe_2O_3 and binary oxides such as $KNbO_3$ and $BaTiO_3$ (Komarneni et al., 1992). Also, a new layered alumina phase was synthesized, which can be intercalated with ethylene glycol. The system is controlled by pressure (which determines the temperature), and other variables such as time, concentration of metal solution, pH, etc., that are used to control the final composition, crystal size, morphology, and level of agglomeration. It was found that microwave-hydrothermal synthesis enhances the apparent kinetics of crystallization of the various oxides by one or two orders or magnitude over that of conventional (Parr bombs) methods. In some cases, the conventional methods, in addition to being much slower, did not lead to the crystallization of a pure oxide phase, as shown in Table 5-4 for TiO_2.

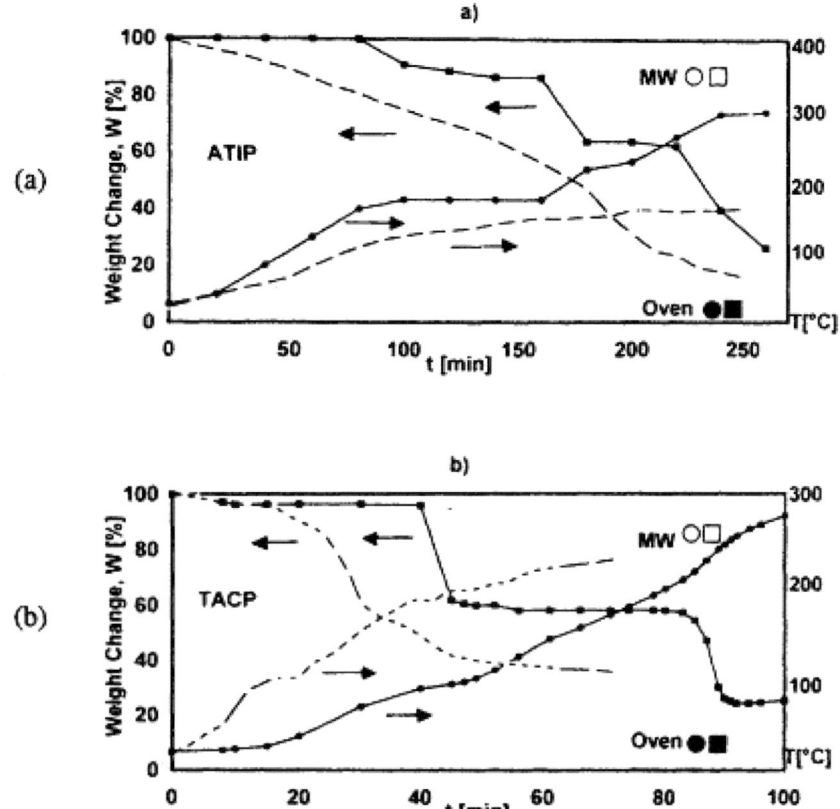

FIGURE 5-2
Oven versus microwave pyrolysis of grin alumina infiltrated with (a) $Al(O\text{-}i\text{-}C_3H_7)_3$ (ATIP), and with (b) $Ti(O_2C_5H_7)_2(OC_3H_7)_2$ (TACP) (Willert-Porada, 1993).

TABLE 5-4 X-ray Diffraction Analyses of Titania Powders Produced by Microwave-Hydrothermal and Conventional Hydrothermal Techniques (From Komarneni et al., 1992)

Concentration (M)		Temperature (°C)	Duration (h)	Reaction Products in order of abundance as determined by x-ray diffraction
$TiCl_4$	HCl			
Microwave-hydrothermal				
0.5	1	164*	0.5	Rutile
0.5	1	164*	1	Rutile
0.5	1	164*	2	Rutile
Conventional-hydrothermal				
0.5	1	164	2	Anatase, small amount of rutile
0.5	1	164	24	Rutile, small amount of anatase
0.5	1	164	72	Rutile, small amount of anatase and amorphous

* Based on autogenous pressure of 200 psi.

In summary, the microwave synthesis of powders is a new era of processing and provides many opportunities for future developments. Table 5-5 presents some of the ceramic powders that have been synthesized using microwave energy.

Powder Treatment

Drying

Because of the strong tendency of moisture to absorb microwaves, and because of the internal (volumetric) deposition of energy, polymeric, ceramic, and other powders can be efficiently dried, and desired residual moisture contents can be precisely controlled. This is an area where microwave processing is in common use (Chabinsky and Eves, 1986; Metaxas and Meredith, 1983).

Calcining

Powdered mixes have been calcined (reactively sintered) using microwaves; the mixes react to form compounds such as $BaTiO_3$ and $NaTiO_3$ (Oda and Balboa, 1988), $Al_6Si_2O_{13}$ (mullite; Willert-Porada et al., 1992) and Al_2TiO_5 (Boch et al., 1992).

TABLE 5-5 Ceramic Powders Synthesized by Microwave Heating

Composition	Process	Composition	Process
Oxide		**Nonoxide**	
Al_2O_3	Solution[1] Pyrolysis[2] Hydrothermal[3]	CrB	Solid-State[7]
		Fe_2B	Solid-State[7]
Fe_2O_3	Solution[1] Hydrothermal[3]	ZrB_2	Solid-State[7]
TiO_2	Solution[1]	AlN	Gas-Phase[8]
Ti_2O_3	Gas Solid[4]	Si_3N_4	Gas-Phase[8]
ZrO_2	Solution[5] Pyrolysis[2] Hydrothermal[3]	SiC	Gas-Phase[8] Solid-State[9,10]
$MgAl_2O_4$	Copyrolysis[2]	TiC	Gas-Phase[8] Solid-State[9,10]
$Al_6Si_2O_{13}$	Sol-gel[6] Copyrolysis[2]	NbC	
$CuAlO_2$	Copyrolysis[2]	TaC	Gas-Phase[8] Solid-State[9,10]
$BaTiO_3$	Sol-gel[1] Hydrothermal[3]	**Composite**	
$YBaCu_3O_{7-x}$	Solution[1] Solid-State[7]	$Al_2O_3/ZrO_2/Y_2O_3$	Solution[5]
$Mn_{0.5}Zr_{0.4}Fe_2O_4$	Solution[1]	SiC/SiO_2	Particle + Coating Pyrolysis[2]
$Mn_{0.6}Zr_{0.4}Fe_2O_4$	Solution[1]	TiC/TiO_2	Particle + Coating Pyrolysis[2]
KVO_3	Solid-State[7]	ZrC/ZrO_2	Particle + Coating Pyrolysis[2]
$CuFe_2O_4$	Solid-State[7]	ZrC/SiC	Particle + Coating Pyrolysis[2]
$BaWO_4$	Solid-State[7]	BN/ZrO_2	Particle + Coating Pyrolysis[2]
$La_{1.85}Sr_{0.15}CuO_4$	Solid-State[7]	SiC/ZrO_2	Particle + Coating Pyrolysis[2]
		Al_2O_3/ZrO_2	Copyrolysis[2]
		Al_2O_3/CuO	Copyrolysis[2]

[1] Kladnig and Horn, 1990
[2] Willert-Porada et al., 1992
[3] Komarneni et al., 1992
[4] Bullard and Lynch, 1992
[5] Vollath et al., 1992
[6] Komameni et al., 1988
[7] Mingos and Baghurst, 1992
[8] Singh et al., 1991
[9] Kumar et al., 1991
[10] Kozuka and MacKenzie, 1991

Enhancing Microwave Absorption

As mentioned in the previous section, many electrically insulating materials, such as oxides, are transparent to microwaves at room temperature. Powders of these materials can be made to couple readily by the addition and mixing of polar liquids or conducting particles. Many refractory oxides, such as alumina, mullite, zircon, MgO, or Si_3N_4, have been made to couple effectively with microwaves by the addition of electroconductive particles of SiC, Si, Mg, FeSi, and Cr_2O_3 (Nishitani, 1979). Oxides of Al_2O_3, SiO_2 and MgO have also been effectively heated by the addition of lossy materials such as Fe_3O_4, MnO_2, NiO, and calcium aluminate (Sutton and Johnson, 1980). Mixtures of conducting powders, such as Nb, TaC, SiC, $MoSi_2$, Cu, and Fe, and insulators such as ZrO_2, Y_2O_3, and Al_2O_3 have coupled well with microwaves (Sutton, 1989). Various materials in solution (zirconium oxynitrate, aluminum nitrate, and yttrium nitrate) that are good couplers have also been added to enhance microwave absorption of powdered insulating oxides (Sutton, 1989).

Powder Consolidation/Shaping

Reaction Sintering

Reaction sintering (or bonding) of oxides and silicon nitride using microwaves has been investigated. In the case of silicon nitride, porous powder compacts of silicon have been reacted with nitrogen at elevated temperatures of 1150—1450 °C (Kiggans et al., 1991). The advantages of this promising process are discussed in more detail in a later section of this report.

Melting

Microwaves have been used to melt powdered materials to form coatings on various substrates by using focused millimeter-wavelength beams. Because of their shorter penetration characteristics, these beams have been used to selectively heat and fuse pore-free coatings, such as Al_2O_3, on lower-melting refractory substrates (Sklyarevich and Decker, 1991; Sklyarevich et al., 1992).

Ignition

Since microwaves create volumetric heating, they have been used to initiate internal ignition in mixtures of exothermic powder compacts (Ahmad et al., 1991). This provides thermal gradients and combustion fronts that move in directions opposite to those in powder compacts that are ignited by conventional (external) methods. The reverse thermal gradients and reaction fronts may enable the synthesis of new and unique structures, composition gradients, and improved properties for this class of materials, which are produced by self-propagating high-temperature synthesis processing (Ahmad et al., 1991; Dalton et al., 1990).

Potential Advantageous Applications of Microwave Heating to Ceramics

Due to the strong absorption of microwaves by water, microwave drying of ceramics has been successful for both powders and bulk materials (Smith, 1974). While the high cost of microwave energy makes microwave drying inefficient at high moisture contents, at low moisture contents (less than 5 percent) the removal of water using conventional processes becomes inefficient, making microwave processes more competitive (Sutton, 1989). A hybrid system with both conventional and microwave heat sources may be the best solution for many drying applications (R. D. Smith, 1991). Due to the depth Of penetration, microwave drying is especially promising for removing low moisture contents from thick sections, including foundry cores (Schroeder and Hackett, 1971; Valentine, 1973), ladle linings (Ochiai et al., 1981), and plaster molds (Valentine, 1973; 1977).

Since temperature gradients are a given with microwave heating, processing schemes that take advantage of the temperature gradients may be attractive. Work is already under way in some of these areas, including microwave-assisted chemical vapor infiltration (CVI) (Evans and Gupta, 1991; Day et al., 1993) and microwave-assisted reaction-bonded silicon nitride (RBSN), (Tiegs et al., 1991; Kiggans et al., 1991; Kiggans and Tiegs, 1992; Thomas et al., 1993a, b).

In both CVI and RBSN, temperature-dependent chemical reactions take place to produce the ceramic material. The rate of reaction also depends upon the concentrations of reactants and any product species in the gas phase. Thus, isothermal processing results in preferential reaction at the surface, where the concentration of reactants is maximum. Moreover, the reaction in both CVI and RBSN tends to either seal the surface or otherwise dramatically inhibit gaseous diffusion, resulting in unreacted or uninfiltrated interior regions. Microwave heating in these cases allows the reaction to take place preferentially in the interior and work its way toward the surface, providing higher final density and greater percent conversion in the two processes. Significantly sized specimens have been successfully nitrided and subsequently sintered in a well-insulated microwave system (Tiegs et al., 1991; Kiggans et al., 1991; Kiggans and Tiegs, 1992). Thomas et al. (1993a, b) utilized a single-mode cavity to nitride disks and rods and demonstrated superior conversion than conventional processing. Finally, Day et al. (1993) used microwave-heating-assisted chemical vapor infiltration to make SiC/SiC and Al_2O_3/Al_2O_3 ceramic-matrix ceramic-fiber composites.

As described earlier in this section, the application of microwaves in ceramic powder synthesis offers some unique benefits, especially with respect to producing particles Of submicron (nano) size with controlled compositions. Extremely fine (less than 1 µm diameter) powders can be produced by controlled reactions in sol-gel processing, gas-phase synthesis, solution evaporation/decomposition, or hydrothermal reactions.

There is a growing interest in the use of microwaves to join ceramics (Silberglitt et al., 1993). Rapid, homogeneous joining can be accomplished using selective microwave heating in either a single-mode applicator (Palaith et al., 1988; Fukushima et al., 1990), by focussing the field at the interface, or a multimode (hybrid) applicator, by using susceptors (Al-Assafi and

Clark, 1992) and by using bonding agents with higher loss than the base material (Yiin et al., 1991; Yu et al., 1991). These studies have only shown the feasibility of joining processes. Applicator design to move the processes to production scale on more complex joints is required (Silberglitt et al., 1993).

Successful industrial implementation of microwave processing depends in large measure upon continuous processing schemes in which parts pass through the microwave cavity. Hybrid heating schemes may find important usage in this regard. In any event, application of microwave processing will probably be limited to materials that do not show a large temperature dependence of the dielectric loss factor and thus are susceptible to thermal runaway. Significant effort must be directed toward applicator design, specifically addressing the issue of openings in the microwave applicator for introduction and removal of the parts.

POLYMERS AND POLYMER-MATRIX COMPOSITES

Polymer Processing

There are increasing demands across broad product lines for new polymeric materials and processes that are cost-effective and environmentally safe. Over the past twenty years, research in the area of microwave processing has shown some potential advantages in the ability not only to process polymers at lower cost but to fabricate new materials and composites that may not be possible using conventional thermal treatments.

One of the first industrial applications of microwave radiation for the processing of polymeric materials was the vulcanization of rubber in the tire industry during the 1960s, with commercial application beginning late in that decade (Chabinsky, 1983a, b; Schwartz et al., 1975). The principal mechanism of coupling of the microwave radiation to the material occurred via carbon black fillers already present in many rubber formulations. Since different grades of carbon black had different coupling characteristics (Ippen, 1971), rubber compounders learned to control the heating patterns throughout the multilayered product through variation of carbon grade and concentration. Application of this processing technology was limited due to the nonuniformity of the microwave curing ovens that were available at that time and thermal runaway attributable to increases in dielectric loss with increasing temperature.

The importance of increased throughput and reduced operating costs, along with advances in microwave equipment, fueled a resurgence in robber processing in the 1980s. Microwave vulcanization of extruded robber weather stripping for the automotive and construction industries has found commercial application, with over 600 installations worldwide (Krieger, 1992). Microwave processing offered rubber processors significant advantages over conventional processing, including improved product uniformity; reduced extrusion-line length; reduced scrap; improved process control and automation; continuous vulcanization rather than conventional batch processes; and improved cleanliness and environmental compatibility compared with steam autoclaves, hot air, salt bath, or fluid bed heating processes.

MICROWAVE APPLICATIONS

There is significant interest in applying this technology to the processing of high-performance, high-cost materials, such as reinforced composite materials including carbon, glass, and ceramic-fiber reinforcement; ceramics; and high-temperature polymers.

Mechanism of Microwave Coupling in Polymers

The principal mechanism of microwave absorption in a polymer is the reorientation of dipoles in the imposed electric field. As in a home microwave, the materials with the greatest dipole mobilities will exhibit the most efficient coupling. Microwave heating, therefore, will couple most efficiently with the strongest dipole in a system and has the potential to selectively heat polar polymers in mixtures. The efficiency of microwave coupling with polymer materials is dependent on the dipole strength, its mobility and mass, and the matrix state of the dipole (Metaxus and Meredith, 1983). Microwave coupling to a given dipole will be greater in a liquid, less in a rubber, and even further reduced in a glassy or crystalline polymer.

Polymer dielectric constants can vary during a processing cycle or if a phase change occurs as temperature varies, solvent is removed, and the reaction proceeds changing the type and concentration of dipoles. Generally, several distinct dielectric relaxation processes are present in a solid polymeric material. This is shown in Figure 5-3, which is a scan of dielectric

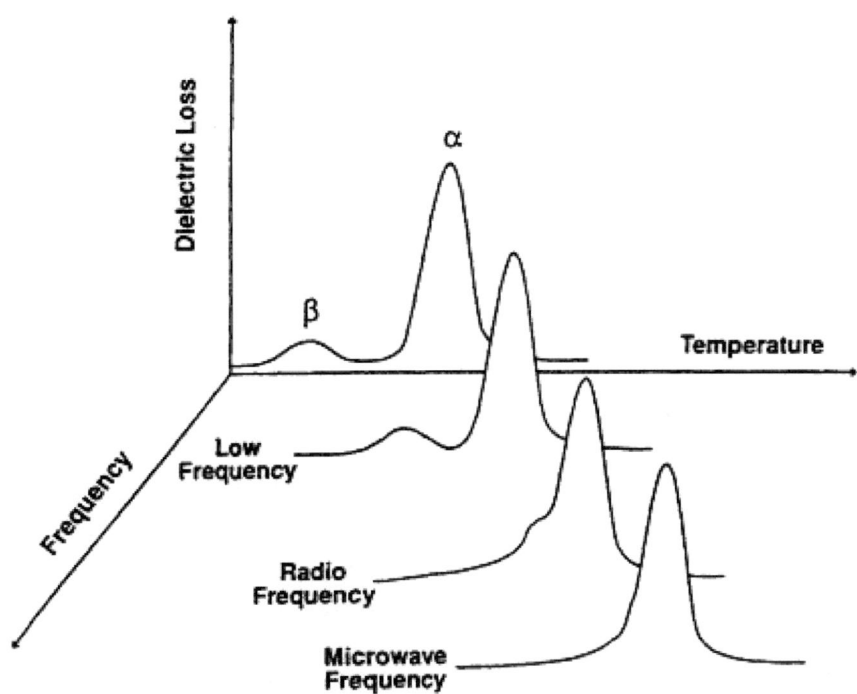

FIGURE 5-3
Schematic dielectric loss versus temperature and frequency for polymer materials (from Chen et al., 1991).

loss at constant frequency as a function of temperature. Similar relaxation processes are observed in dynamic mechanical properties of polymers, with analogous dispersions in real and imaginary components of viscoelastic response (Ward and Chen, 1992). An excellent review of the dielectric properties of polymers at microwave frequencies is presented by Bur (1985).

Polymers for Microwave Applications

Not all polymer materials are suitable for microwave processing. However, many polymers contain groups that form strong dipoles (e.g., epoxy, hydroxyl, amino, cyanate, etc.). Microwave processing can be used over a broad range of polymers and products, including thermoplastic and thermosetting resins, rubber, and composites.

Initially, thermosetting polymers are low-viscosity liquids that can flow into a mold or around fibers. During processing, thermosets react to increase molecular weight and viscosity, eventually becoming highly cross-linked, insoluble, infusible materials. Thermoset cure processes consist of three basic steps: (1) preheating of the components; (2) reaction, producing the corresponding exotherm; and (3) cooling of the cured materials (Van and Gourdenne, 1987). Permittivity and dielectric loss factor of thermosets generally increase with temperature and decrease with extent of cure (Jow et al, 1988). These polymers tend to be efficient absorbers of microwave radiation initially, with ϵ'' increasing as the resin is heated. As the cure reaction progresses, the temperature may be difficult to control due to the additional heat input caused by the exothermic reaction.

Thermoplastics are fully polymerized materials that melt and flow upon application of heat. They are processed well above their glass transition temperatures or melting points (if the material is semicrystalline) to reduce the melt viscosity and allow flow and to promote adhesion. High-performance, semicrystalline thermoplastic polymers, such as Polyetheretherketone (PEEK), can be difficult to heat using microwaves until a critical temperature is reached, where ϵ'', and therefore the heating rate, increases significantly (Chen et al., 1989). This critical temperature is related to increased molecular mobility but may not be the same as the glass transition temperature of the polymer. The crystallinity of these materials is important; amorphous polymers heat more effectively than semicrystalline polymers (DeMeuse, 1992).

Functionally terminated thermoplastics combine the toughness of thermoplastics with the ease of processing and the creep resistance and solvent resistance of thermosets. These materials undergo a combination of thermoset and thermoplastic processing with the initial heating reducing viscosity and improving flow and ultimately reaction providing a cross-linked network. Microwave processing of functionally terminated thermoplastics offers advantages over conventional processing, particularly in reducing the processing time (Hedrick et al., 1989). One of the challenges in the microwave processing of these polymers is that the processing temperature is often very close to its thermal degradation temperature, making temperature control crucial. If the temperature is too high, the polymer undergoes undesirable cross-linking, scission, and oxidation, which can cause significant changes in the mechanical and optical properties of the material. This behavior places a strict requirement on the microwave system to provide very uniform temperature distributions throughout the part being processed and careful control of the temperature of the part.

Although the polymer systems that are candidates for microwave processing are typically not conductive, particles and fibers that are conductive, or have dielectric properties significantly different from the matrix polymer, may be included to aid processing or to modify the mechanical, physical, or optical properties. The presence of these inclusions can strongly influence the way in which the composite material interacts with the microwave radiation. Conductors also modify the electric field pattern in and around the composite, potentially resulting in very different heating profiles than with the neat resin. Some examples of these conductive additives include carbon black (used extensively in rubber formulation); carbon or metal fibers; and metal flakes, spheres, or needles with sizes ranging from 0.1 to 100 µm. Although the final composite is not necessarily conductive, the surfaces of the conducting inclusions interact strongly with the microwave radiation. The effect of conductive additives on microwave heating and skin depth of the composite depends on the size, shape, concentration and electrical resistivity of the inclusions and their distribution in the matrix (Lagarkov et al., 1992).

The presence of conducting fillers may inhibit microwave heating by decreasing skin depth. However, by controlling the nature, orientation, and concentration of the fillers, the microwave response of the material can be tailored over a broad range. For example, carbon fibers have a relatively high resistivity and heat the surrounding matrix very effectively; the thermal profile has a maximum at the surface of the fibers. This preferential heating has been shown to provide an enhancement of the interfacial adhesion between the fibers and the matrix resin (Agrawal and Drzal, 1989) and a subsequent improvement in the fracture properties of microwave-processed composite materials. Preferential heating of conducting fillers has also been utilized in the joining of polymers and polymeric composites (Varadan, et al., 1990). Baziard and Gourdenne (1988a, b) report an increased rate of cross-linking in a composite system of an aluminum powder and epoxy resin. The rate of cross-linking is attributed to the higher dielectric loss due to the presence of the filler. Similar results for carbon-black filled epoxy resin systems have been reported (Bouazizi and Gourdenne, 1988).

Nonconductive additives such as glass fibers and nonconducting metal oxides which are used as pigments (e.g., titanium dioxide), can also influence composite properties through preferential heating mechanisms, depending on their dielectric properties.

Enhanced Reaction Kinetics

In addition to the efficient coupling of microwave energy in polar materials and significant depth of penetration, nonthermal "microwave effects," including accelerated apparent kinetics (Lewis et al., 1992, 1987, 1988; Hedrick et al., 1989), retarded kinetics (Mijovic and Wijaya, 1990), and dependencies of the heating rate (Gourdenne, 1992; Chan and Gourdenne, 1992) and structure (Thuillier et al., 1986) of the cured polymer structures formed on the pulse repetition frequency have been reported. The most prevalent reports of microwave effects have been acceleration of reaction rates. There have also been reports in which no effect of the radiation on the kinetics was observed (Mijovic et al., 1992a, b; Jullien and Petit, 1992; Jordan et al., 1992; Mijovic et al, 1991).

One of the difficulties in the comparison and rationalization of these effects is that the experimental conditions and the materials have differed from group to group. A number of the reports of microwave effects are for the curing of epoxy resins and simply measure conversion with time (Boey et al., 1992). Unfortunately, it is difficult to analyze these data further, since the curing reaction can behave autocatalytically. Even within the general class of epoxy resins on which a large amount of work has been performed, the reactivity can vary more than an order of magnitude depending on the resin constituents and formulations. Furthermore, as the reaction progresses, molecular weight and cross-link density increase, limiting molecular mobility (which limits reaction rate) and making comparison of reaction kinetics difficult, especially at high conversions. A meaningful kinetic analysis must account for the development of network structure and the resulting reduction in mobility of reactive groups (Wingard and Beatty, 1990; Woo and Seferis, 1990).

Two general observations that can be made are that (1) slower-reacting systems tend to show a greater effect under microwave radiation than faster-reacting systems and (2) the magnitude of the observed effect decreases as the temperature of the reaction is increased. The manner in which temperature is measured and controlled is critical in kinetic analysis. The challenges associated with temperature measurement in a microwave field are discussed in Chapter 3.

A number of problems associated with kinetic analysis of a reacting system were avoided in a study of a solution imidization reaction. This reaction followed first-order kinetics, and the reactant and product remained in solution throughout the reaction (Lewis et al., 1992). Isothermal conditions were maintained by varying the microwave power or detuning the applicator. An 18- to 35-fold enhancement in the reaction rate was reported over the temperature range studied. The enhanced reaction rate corresponded to a reduction in the activation energy for the reaction from 105 kJ/mole to 55 kJ/mole.

A proposed mechanism for the "microwave effect" in polymers suggests a nonequilibrium, nonuniform energy distribution on the molecular level, which results in certain dipoles having a greater energy than the "average" energy of adjacent groups (Lewis et al., 1988, 1992). For the solution imidization of a poly(amic acid), this increased energy corresponded to an increase in an effective temperature of the reacting groups of approximately 50 °C over the bulk temperature. The energy couples directly with a reactive polar group in this system and dissipates through adjacent groups by the usual mechanisms. However, if the energy is absorbed faster than it is transferred, at least initially, there will be a nonuniformity present. This mechanism is consistent with some of the recent pulsed-radiation studies in that the rate of energy transfer along the chain may be related to chain relaxations that occur on a similar time scale to the pulse repetition frequency.

Because of the range of materials studied, differences in temperature control and measurement methods, and variations in microwave applicators, based on available data it is impossible to determine the effect that microwave processing has on reaction kinetics. Consistent, controlled experiments, with careful measurement and control of temperature, that account for variations in resin chemistry and changes in reaction mechanisms during cure, are needed to investigate nonthermal microwave effects.

Polymer-Matrix Composites

High-performance polymeric composites, reinforced with carbon, glass, or aramid fibers, have been effectively used by the aerospace and electronics industries in applications requiring light weight, high specific strength and stiffness, corrosion and chemical resistance, and tailorable thermal-expansion coefficients. The dielectric properties of glass or high-performance, polymeric fiber-reinforced composites have made them attractive for printed circuit boards and in aviation, marine, and land-based systems radome applications.

More-general application of polymeric composites has been hindered by their high cost of orientation (layup) and forming (molding and curing) processes. Innovative processing, including automated lamination, rapid consolidation and curing, and out-of-autoclave processing, is being pursued in an attempt to reduce the costs associated with processing. Microwave processing shows promise for rapid, nonautoclave processing of composite structures.

The processing of very thick cross-section parts using conventional processing requires complex cure schedules with very slow thermal ramp rates and isothermal holds to control overheating due to cure reaction exotherms and poor thermal conductivity. Because of microwave penetration and rapid, even heating characteristics, thick composites were initially targeted as ideal applications for microwave processing. Early studies (Lee and Springer, 1984a, b) indicated that, while microwave curing of composites in wave-guide applicators was feasible, materials with conducting (carbon) fibers would be limited to unidirectional composites with less than about 32 plies (approximately 7—8 mm thick) due to the high reflectivity of the fibers and, hence, poor penetration depth of the radiation into the composite.

Tunable, single-mode resonant cavity applicators with feedback controls to allow the resonant frequency to be changed as material properties vary during processing have been developed to allow more-efficient coupling with composites (Asmussen et al., 1987, Asmussen, 1992). Much of the work accomplished in polymer and composite processing has utilized this type of cavity applicator.

The feasibility of curing thick cross-plied carbon fiber composites was shown when 36-and 72- ply composites were successfully cured using a single-mode resonant cavity (Wei et al., 1991). Heating was controlled through feedback on/off switching of microwave power based on sample temperature as measured using a fluoro-optic probe. The characteristic temperature excursion resulting from the exothermic reaction during epoxy cure was eliminated by using a pulsed system that allowed a higher temperature cure without thermal degradation (Jow et al., 1989; Jow, 1988). The mechanical properties of microwave-processed glass/vinyl ester composites were shown to be at least equivalent to those of conventionally processed materials, with indication that some property enhancement attributable to reduced void content occurred (Ramakrishna et al., 1993). Increased adhesion and improved mechanical properties at the fiber/matrix interface were observed for carbon-fiber composites due to preferential heating at the conductive fiber surface (Drzal et al., 1991). Although this process works well for flat parts, tuning of a single-mode cavity containing complex or large parts to provide uniform heating has not yet been accomplished (Fellows et al., 1993).

A tunable single-mode applicator was used to heat carbon-fiber reinforced PEEK thermoplastic (Lind et al., 1991). Enough power was absorbed to rapidly heat the PEEK matrix to melt temperatures so that it could be bonded to a consolidated laminate. Based on these results, an applicator was designed and preliminary concepts were developed for an automated tape placement process for fabrication of composite parts (Figure 5-4). Feedback controls to adjust cavity resonance to account for panel curvature are required for scaling.

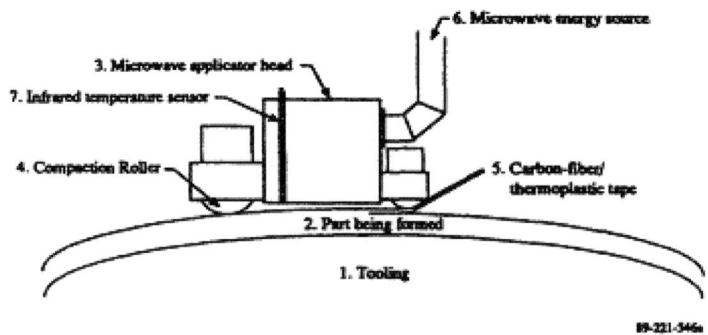

FIGURE 5-4
Concept drawing showing use of a microwave applicator in a tape-placement part-forming device (from Lind et al., 1991).

Potential Applications of Microwaves to Polymers and Composites

Although various studies have claimed that the microwave curing processes can have cost advantages over conventional processes (Simonian, 1979; Chabinsky, 1988; Akyel and Bilgen, 1989), microwave processing of polymers has not found widespread industrial application. However, there are polymer processes that are particularly promising for industrial application.

An area where microwave processing has shown promise is composite pultrusion. In pultrusion, a polymeric composite preform is pulled through a heated die, where the shape is molded and the matrix cured. In conventional processes, the processing chamber consists of a heated die, which is quite long due to the slow heat transfer to the polymer matrix and relatively long cure times. A single-mode resonance cavity has been used to rapidly heat the part using microwave radiation in a significantly shorter process chamber, resulting in less force required to pull the fiber bundle through the die (Methven and Ghaffairyan, 1992). Since the part configuration that the applicator sees is fixed for each shape, process control should be relatively simple.

In a process analogous to pultrusion, polymeric fibers are drawn through heated dies to increase their axial strength and stiffness through polymer chain orientation. When microwave radiation was utilized for drawing fibers, it was shown that the draw ratio could be increased from approximately 20:1 to 35:1 with a corresponding increase in the modulus from 35—40 GPa for conventional processing to 55—60 GPa for poly(oxymethylene) (Nakagawa et al., 1983; Takeuchi et al., 1985; Nakagawa et al., 1985), with similar results for other polymers (Amano and Nakagawa, 1987a, b). The significantly superior mechanical properties of microwave ultradrawn poly(oxymethylene) fibers over conventionally processed fibers (Nakagawa et al, 1983) were due to the increased orientation of the polymer chain in the fiber direction.

Because microwaves will couple selectively with materials that contain polar functionalities, it is possible to combine the efficiency and uniformity of heating with the selectivity of materials and accelerate and improve adhesion by coupling microwave energy directly into the adhesion interface. Recently, it has been shown that an intrinsically conducting organic polymer "self-heats" when it is exposed to electromagnetic radiation from a microwave, dielectric, or induction source or when a current is passed through it. The high dielectric loss tangent of a conducting polymer such as polyaniline (loss tangent greater than or equal to 10^{-1} at 6.5 GHz) is responsible for its microwave heating (Epstein et al., 1993). This heat is sufficient to locally melt and weld adjoining thermoplastic parts or cure thermoset polymers, but it does not heat the entire structure, which can result in softening or distortion. This phenomenon can be used to fabricate strong joints of plastics or composites either with each other or with metals. Extensive work has been done on the microwave welding of high density polyethylene (HDPE) using conductive gaskets made from a blend of HDPE and conducting polyaniline (Wu and Benatar, 1992). Under optimum welding conditions, the microwave-welded joint had a tensile strength equal to that of the bulk material.

MICROWAVE PLASMA PROCESSING OF MATERIALS

Microwave excitation readily forms plasmas at reduced gas pressures and, under some circumstances, at pressures in excess of 1 atm. Microwave plasmas are being utilized extensively for various applications in microelectronic processing, including deposition and etching for diamond film deposition; for surface modification; and, on an experimental basis, for sintering of ceramics. An important application of microwave plasmas, analytical spectroscopy, is outside the scope of this study.

Plasmas interact with surfaces in one of two ways beyond simply providing thermal energy for heating. Atomic or ionic species in the plasma may react with the substrate to form volatile constituents (etching), or species in the plasma may react to form solid materials, which are deposited on the substrate (plasma-enhanced chemical vapor deposition). Plasma surface modification processes may involve either of these interactions.

Microwave plasmas are generated in single-or multimode cavities, electron cyclotron resonance cavities, and coaxial torches. Coaxial torches find little use in materials processing. Microwave plasmas, in contrast to parallel plate RF plasmas, do not involve electrodes in contact with the plasma. This avoids contamination arising from sputtering from the electrodes. The specimen may be in direct contact with the plasma, or the effluent of the plasma may be utilized in the processing.

There are significant differences between microwave plasmas and the more common parallel-plate RF plasmas that are used for microelectronics processing. In RF plasmas, one or both of the electrode plates is excited at radio frequency, typically 13.56 MHz. A large DC bias is developed between the plasma and the electrode on which the specimen rests, causing bombardment of the specimen with directed high-energy ions. This phenomenon is utilized in the reactive ion etching RF systems. In a microwave plasma, a much smaller bias is developed between the plasma and the specimen than in RF plasmas. In addition, the degree of ionization is greater in the microwave plasma. These characteristics have significant consequences in plasma processing. Depending on the process, the differences may be an issue in deciding whether to use a microwave plasma. An excellent review of microwave plasmas has appeared recently (Moisan and Pelletier, 1992).

The current literature on microwave plasma processing is heavily dominated by reports on diamond film formation. The growth of diamond films requires an abundance of atomic hydrogen, which etches graphitic nuclei in the deposit and leaves the diamond-like nuclei to grow. Plasmas generated by any means are, in general, good sources of this species. There are certain perceived advantages of microwave plasmas over other diamond film-forming methods. Cited examples include stability and reproducibility of the plasma, energy efficiency, availability of inexpensive magnetrons, and potential for scaling to larger sizes (NRC, 1990). A further advantage is that the microwave plasma can heat the substrate to the temperature required for good deposition conditions (greater than 500 °C).

Microwave plasma processing has had a major impact in microelectronics device processing, where it is a mature art. A state-of-the-art review listed microwave plasma processing as a key technology that was sufficiently developed for imminent implementation in industry (NRC, 1986). The two major applications are plasma-enhanced chemical vapor deposition and etching, which includes the possibility of high-resolution etching of silicon (Moisan and Pelletier, 1992).

Deposition

Microwave plasma deposited materials include silicon films, which are amorphous or polycrystalline depending upon the substrate temperature, and silicon oxide and nitride. In addition, silicon can be oxidized to form silicon oxide films. The primary advantage of microwave plasma-enhanced chemical vapor deposition is reduction in radiation damage compared with conventional RF plasma chemical vapor deposition. This is because the microwave discharge results in a lower acceleration potential between the plasma and the substrate. The electron cyclotron resonance plasma technique is particularly useful in depositing silicon oxide and silicon nitride films on silicon for device processing. Films deposited at temperatures less than 150 °C have chemical and physical properties equivalent to films deposited at 900 °C using conventional chemical vapor deposition processes, and the low-energy ion bombardment does not damage the substrate. Similarly, silicon oxide films grown on silicon appear comparable to those grown by conventional thermal oxidation at 1000 °C. Proper design of equipment, including positioning of feedstock injection, is important to avoid unwanted depositions on the walls of the reactor or other places.

The second major application of microwave plasmas is etching in electronic device processing. The principal advantage is that the microwave plasmas are more selective between photoresist and the underlying material. The second advantage is the lower intensity of radiation damage in reactive ion etching compared with conventional plasma etching because of the lower acceleration potential for ions. Finally, microwave plasma etching is reported to give highly anisotropic etching, although an RF bias is usually required to achieve the desired level of anisotropic etching (Moisan and Pelletier, 1992).

An RF bias to a microwave plasma not only increases the directionality of the etching, it also increases the rate of etching. Thus the microwave plasma is more selective than the RF plasma, whereas the RF plasma provides better directionality, so a combination of the two is required to obtain the desirable degree of both selectivity and directionality.

Surface Treatment

A third area of use of microwave plasmas is in surface treatment, where it has been applied to polymer fibers, as well as in the microelectronics industry. Chemical modification of the surface can be achieved with or without adding reactive components in the plasma. It has been demonstrated that treatment of polyamide fibers in a large microwave plasma system improves the bonding between the fiber and the matrix in composites (Wertheimer and Schreiber, 1981). This results in a dramatically different response to mechanical loads, providing for higher strength but at the same time a lower ballistic strength. Fiber mechanical properties can be degraded by the microwave plasma treatment.

Microwave plasmas are used also to promote adhesion of films in microelectronics processing. Advantage is taken here of the lower degree of radiation damage that is achievable with the microwave plasma than with other plasmas. By using a combination of microwave excitation and RF biasing, it is possible to independently control the relative contribution of the chemical component and the physical component (energetic ions, electrons, and photons).

In addition, microwave plasma sources have been used to passivate the surface of GaAs, resulting in superior device properties. The avoidance of direct ion bombardment of the surface was key to the success of this application.

The interactions among the physical and chemical components of a microwave plasma system are numerous and not well understood. Further work of a basic nature is required to better elucidate these interactions. Until then the industry and art probably will be dominated by solutions arrived at by trial and error. One of the aspects that should be explored in more detail is the effect of variable frequency on the chemical and physical processes occurring in the microwave plasma and on interactions with the substrate during deposition, etching, and surface modification.

MINERALS PROCESSING

The minerals and extractive metallurgy industry is a major consumer of energy and contributor to environmental degradation. For instance, about 4 percent of the carbon dioxide emitted to the atmosphere comes from the worldwide extractive metallurgy industry (Forrest and Szekely, 1991). Microwave processing may provide substantial benefits in reducing energy consumption and environmental impact by this industry.

In mineral processing, the extraction of values in an ore from the waste or gangue is an energy intensive and energy inefficient process. According to Walkiewicz et al., (1991) approximately 50—70 percent of the energy used for minerals extraction is consumed during comminution (grinding) and separation. The energy efficiency of conventional grinding is about 1 percent, and most of the energy is wasted in heat generated in the material and equipment.

Microwave processing of ores provides a possible mechanism to induce fractures between the values in the ore and the waste material surrounding it, due to the differential in absorption of microwaves and the differences in thermal expansion among various materials. These differentials induce tensile fractures in the material (Figure 5-5), and as a consequence, substantially reduce the energy required in grinding to separate the values from the waste material.

A limited amount of work has been done in this area, and it is clear that microwaving of certain sulfide and oxide ores does result in fractures along the interface between the values and the waste material. Grindability tests show improved grindability (less energy is required to achieve a given mesh size for the ore) for a series of iron ores. However, it is not clear that the reduction in energy required in grinding will balance or exceed the energy expended in the microwave treatment of the ore. In iron ores the preliminary results indicate a deficit in the energy balance. To justify using microwave processing, it is also necessary to consider wear on grinding mills, cleaner liberation of the values in the ore, and lower chemical emission during the pyrometallurgy and hydrometallurgy processing steps.

Additional research needs to be done to determine the efficiency of coupling the microwave energy to the ore, the effect of particle size on susceptibility to cracking, and the effect on the cracking efficiency of using high power sources (the present work has been limited to maximum of about 3 kW).

MICROWAVE CHEMISTRY

Microwave chemistry is a rapidly growing field that has been gaining attention recently (IMPI, 1992; EPRI, 1993). The effect of microwave processing on chemical reactions or processes touches on most of the application areas emphasized elsewhere in this report. Some of these include ceramic sintering and synthesis, polymer curing, plasma processing, and waste remediation. In this section, applications in analytical and synthetic chemistry and extensions of these applications to the chemical industry are considered.

The most widespread use of microwaves in chemistry is in analytical laboratories. Microwave energy has been used in analytical chemistry since the mid-1970s, primarily for sample preparation. In that time, microwave ovens have become generally accepted tools in the modern analytical laboratory, increasing from a couple hundred units in 1975 to close to 10,000 units in 1992, while the annual expenditures for laboratory microwave systems increased from under $1 million in 1975 to close to $50 million (Neas, 1992a).

Applications span a wide range of sample preparation methods including drying, extractions, acid dissolution, decomposition, and hydrolysis. In these applications, microwave heating has been used as a replacement for conventional heating techniques. In general, analytical chemistry involves time-consuming sample preparation steps to get the samples in a suitable form for analysis.

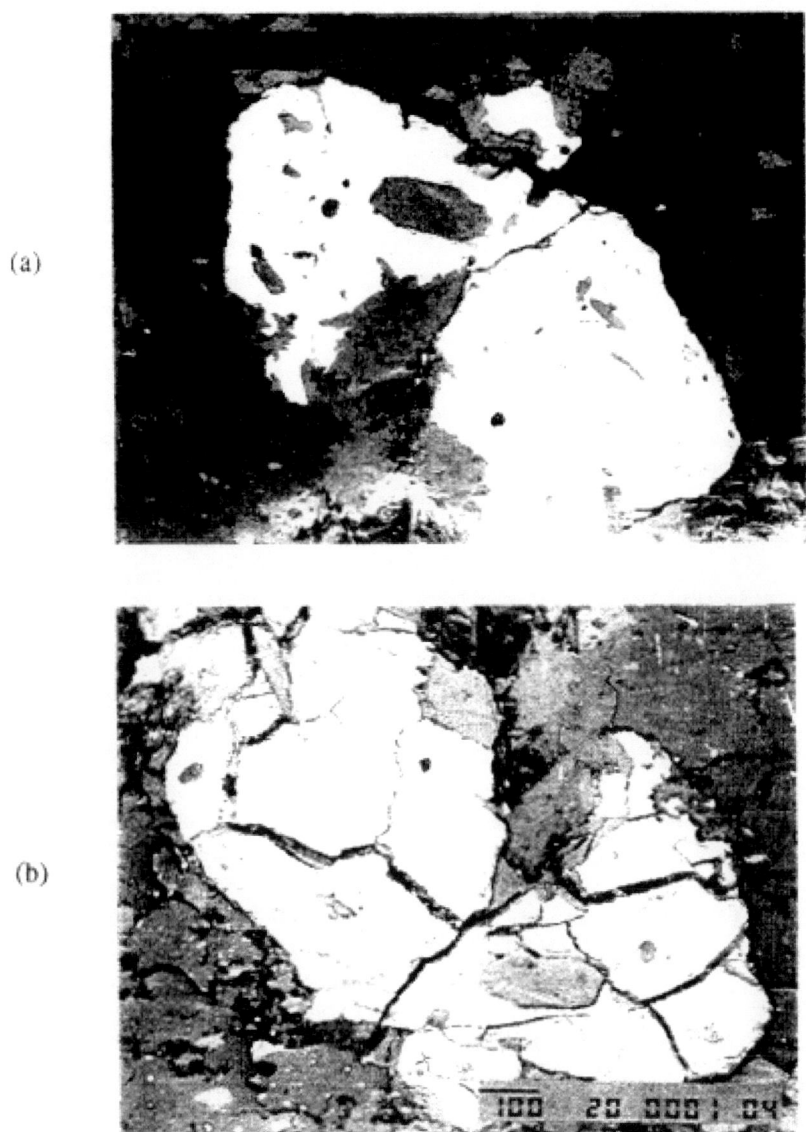

FIGURE 5-5
Photomicrograph of pyrite ore (a) nonmicrowaved and (b) microwaved, showing stress cracking. The light phase is pyrite; and the dark phase is quartz (magnified 100×). (Courtesy of J. Walkiewicz, U.S. Bureau of Mines)

Microwave digestion of materials, such as minerals, oxides, glasses, and alloys, is used in laboratories worldwide to prepare samples for chemical analysis. The decomposition rate of many difficult-to-dissolve materials in closed-reaction vessels is greatly enhanced by using microwave energy; often only a few minutes are required as opposed to the several hours needed for conventional means (Kingston and Jassie, 1988a, b). In addition, volatile elements such as selenium and phosphorous can be quantitatively retained in a sealed vessel using microwave decomposition prior to instrumental analysis (Patterson et al., 1988).

Applications in the chemical laboratory generally use relatively simple ovens and controls. Most of the development work in equipment for these applications has involved improving the existing equipment to extend the operating range and to improve safety and reproducibility. Examples include improved turntables and sample fixtures, pressure vessels fabricated from glass and quartz to allow higher reaction temperatures and pressures than teflon vessels, and optimized pressure relief valves (Baghurst and Mingos, 1992b). The result of these advances has been the development of testing standards that are simple, reproducible, and automatable (Kingston, 1992). A wide variety of organic synthetic reactions have been shown to be enhanced by microwave processing (Bose et al., 1993; Majetich and Hicks, 1993). Using microwave processing, a number of fundamental organic reactions have shown accelerated reaction rates and increased yields over conventional techniques. While these processes have not yet been scaled to production, important advantages have been realized in education, where reactions that took too long to accomplish in a laboratory session using conventional heating can now be completed using microwave heating.

The primary motivation for use of microwave heating has been time savings through rapid heating, rather than any nonthermal effects. Penetrating radiation (and reverse thermal gradients), the ability to superheat polar solvents, and the ability to selectively heat reactive or catalytic compounds were responsible for time savings realized in chemical processes.

Microwave energy penetrates into the interior of the sample without relying on conduction from the surface required in conventional heating methods. This allows the entire sample temperature to be raised rapidly without overheating, and possibly degrading, the surface. Convective heat losses from the surface to the cooler surroundings allow processors to take advantage of reverse thermal gradients.

The reaction temperature of solvent diluents can be raised above the ambient boiling points of the diluents in both closed-and open-reaction vessels (Baghurst and Mingos, 1992a). This allows for significant increases in reaction rates in a variety of applications (Mingos, 1993). Reaction rate enhancements were attributable to Arrhenius rate effects due to increased reaction temperature or selective heating of reactants over diluents. There are no persuasive arguments to support nonthermal reaction enhancements attributable to the use of microwaves (Majetich, 1992; Mingos, 1992).

In closed vessels, the increased vapor pressure over the liquid suppresses further boiling. Microwave superheating of volatile solvents can lead to significant acceleration of chemical processes compared with conventional reflux conditions. The development of microwave transparent glass and quartz reaction vessels and improved pressure-relief valves has been critical in allowing attainment of higher temperatures and pressures than was possible with low-loss teflon vessels (Baghurst and Mingos, 1992b).

In open vessels, most polar solvents have an inherent ability to be heated above their conventional boiling points. This effect has been observed by several researchers (Mingos, 1993; Baghurst and Mingos, 1992a; Neas, 1992b; Majetich, 1992). The phenomenon has been explained using a model of nucleation-limited boiling point (Baghurst and Mingos, 1992a). During a boiling process, bubbles nucleate preferentially at sites (cavities, pits, scratches) on the vessel wall, allowing growth of the vapor phase. With conventional heating, the vessel wall and liquid surface are generally hotter than the bulk. In microwave processes, however, the vessel wall is cooler than the bulk solution due to convective heat losses from the surface, allowing the bulk to attain temperatures above the conventional boiling point before the boiling process commences. As shown in Table 5-7, the nucleation-limited boiling point varies with the solvent and with the ability of the solvent to wet the vessel surface, that is, the more effectively the solvent wets the vessel, the more difficult bubble nucleation becomes.

TABLE 5-7 Nucleation-Limited Boiling Points for a Range of Solvents

Solvent	B.p./(°c)	Nucleation-Limited Boiling Point (°C)	Boiling Point Change (°C)
Water	100	104	4
Ethanol	79	103	24
Methanol	65	84	19
Dichloromethane	40	55	15
Tetrahydrofuran	66	81	15
Acetonitrile	81	107	26
Propan-2-ol	82	100	18
Acetone	56	81	25
Butanol	118	132	14
1.2-Dimethoxyethane	85	106	21
Diglyme	162	175	13
Ethyl acetate	78	95	17
Acetic anhydride	140	155	15
iso-Pentyl alcohol	130	149	19
Butan-2-one	80	97	17
Chlorobenzene	132	150	18
Trichloroethylene	87	108	21
Dimethylformamide	153	170	17
Chlorobutane	78	100	22
iso-Propyl ether	69	84	16

Source: Baghurst and Mingos, 1992a.

By using microwave heating, the processor is able to target compounds with high dielectric loss over less-lossy compounds. This characteristic has been shown to enhance a number of chemical processes, including catalytic reactions utilizing metallic or dielectric catalysts, gas-phase synthesis of metal halides and nitrides, and metal reduction processes (Bond et al., 1992).

The promising future of microwave chemistry to the chemical industry is just beginning to be realized. Advantages in the form of time savings, increased reaction yields, and new processes have been demonstrated in the laboratory using simple multimode ovens. Scaling to widespread production applications requires development of applicators and handling systems to account for increased product throughput, automation, and large reactors.

Microwave processing for small-scale, custom organic synthesis looks promising due to the relatively modest equipment investment, broad applicability to a variety of reactions, and significantly reduced processing times (Bose et al., 1993). The availability of equipment and the long history of use in the laboratory makes these near-term applications low risk. Another area where microwaves can show an advantage is in producing products or intermediates that are needed in small quantities and may be hazardous and expensive to ship, store, and handle (Wan and Koch, 1993).

WASTE PROCESSING AND RECYCLING

The processing of industrial wastes is an area of tremendous promise for the application of microwave energy. The types of industrial waste that have been shown to be amenable to microwave processing, at least at laboratory scale, include hazardous waste (including toxic and radioactive) with high disposal, storage, or treatment cost and nonhazardous waste where the recovery or reuse of a raw material represents a significant cost or energy savings. Waste processing includes treatment or remediation of process wastes, detoxification or consolidation of stored waste, or cleanup of storage or disposal sites.

The application of microwave energy in the processing of industrial waste has seen significant progress in terms of process development and demonstration but limited commercial application. In varying degrees, applications in this area take advantage of unique features of microwave heating: rapid heating, selective coupling with lossy constituents, and reaction steps not possible or practical with other methods.

Process Waste Treatment

Potential applications of process waste treatment include microwave plasma hydrogen sulfide dissociation, detoxification of trichloroethane (TCE) through microwave plasma assisted oxidation, and microwave plasma regeneration of activated carbon.

The dissociation of hydrogen sulfide (H_2S) in a microwave plasma was first described in Soviet literature (Balebanov et al., 1985). The potential for this process is in the refuting industry for the treatment of the sour gas resulting from hydrodesulfurization of hydrocarbon feedstocks. Subsequent work at Argonne National Laboratory has validated the applicability (Harkness et al., 1990) and economic viability (Daniels et al., 1992) of this process in the treatment of refining wastes. A schematic of the microwave dissociation process is shown in Figure 5-6. The attainable conversions, between 40 and 90 percent, were most sensitive to gas flow rate and power. Conversions up to 99 percent are achievable by cycling the residual H_2S back through the process in multiple passes. Work is continuing to scale the process. The economic viability of the process depends on the sale of recovered hydrogen and is sensitive to required dissociation power.

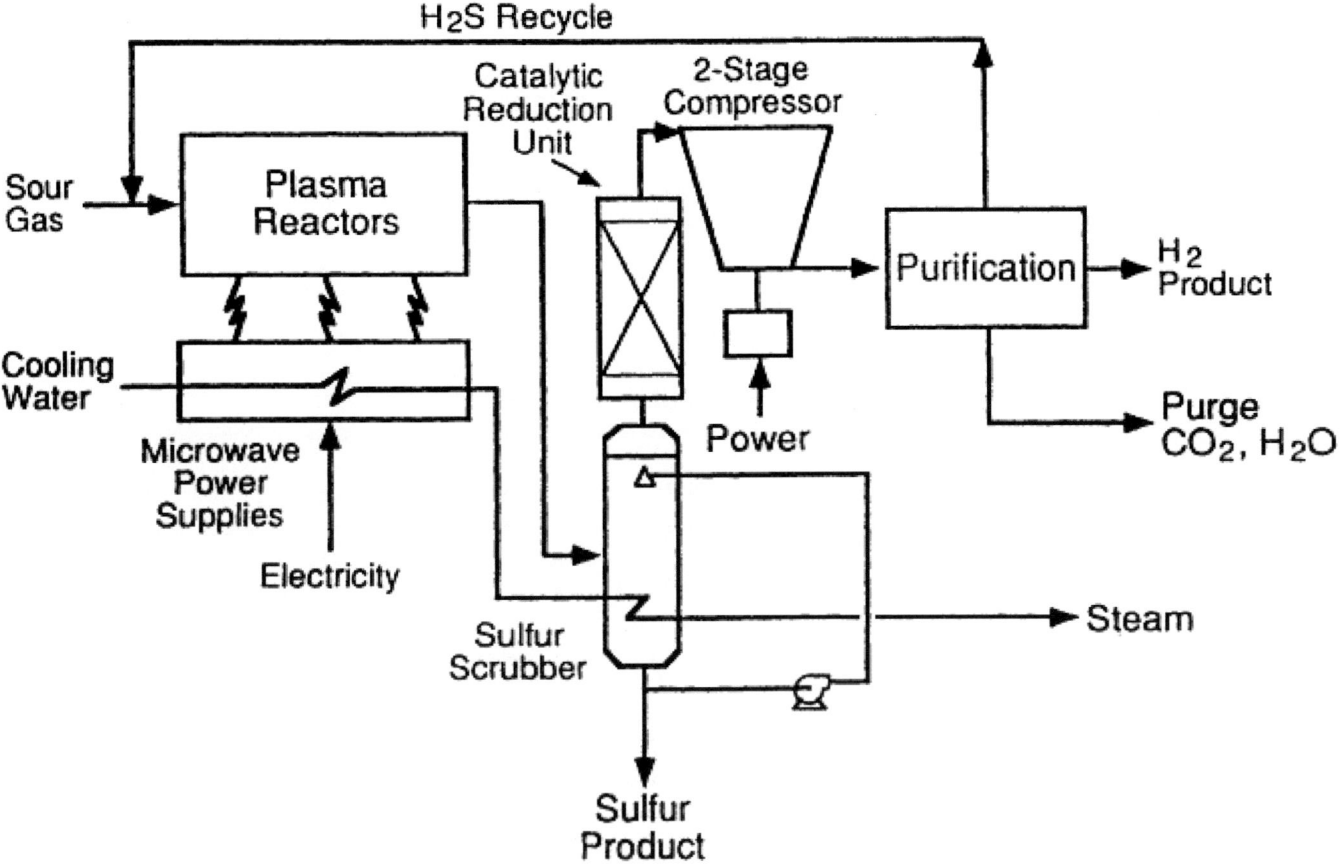

FIGURE 5-6
Hydrogen sulfide waste-treatment process utilizing microwave plasma dissociation. (Courtesy of E. Daniels, Argonne National Laboratory)

TCE oxidation and activated carbon regeneration are accomplished through selective heating of lossy components (SiC or carbon) in a fluidized bed. These processes cause degradation of hazardous organic compounds at significantly lower overall temperatures than conventional heating methods. Additionally, severe corrosion of furnace components caused by the gases released in conventional high-temperature oxidation of chlorinated hydrocarbons is eliminated in the microwave process.

Stored Waste Treatment

Examples of microwave processing of stored waste include "in-can" evaporation of water and consolidation of low-level radioactive waste (Oda et al., 1991; White et al., 1991a). These processes heat low-level sludge wastes in the final storage containers by applying a microwave field using a slotted waveguide applicator. In-can processes use the rapid, selective heating of water possible with microwaves and the portability of microwave equipment to reduce transfer and handling of hazardous materials that is necessary when using conventional drying approaches. This approach need not be limited to radioactive sludge but can be applied to any evaporative treatment of stored materials. Even though, as mentioned earlier, the efficiency of bulk drying using microwaves is questionable, the cost avoidance realized by reducing handling steps may justify the increased energy costs.

Waste-Site Cleanup

The cleanup of contaminated industrial, disposal, and storage sites is a formidable task due to the large number of waste sites and the complex chemistries and remediation requirements involved. Currently, the majority of site cleanup efforts require removal and incineration of the waste. Since removal and transfer of contaminated materials for incineration may represent an unacceptable risk, innovative processes to cleanup contaminated storage and disposal sites are being investigated. Microwave processing shows great promise for site cleanup applications, since microwaves can be applied *in situ*, avoiding costly and risky excavation and transportation, and can target compounds with high dielectric loss for selective heating, for example, moisture in soils (Dauerman, 1992).

Potential applications of microwave processes for cleanup of contaminated sites include removal of volatile organic compounds from soil (George et al., 1991; Windgasse and Dauerman, 1992) and remediation of soils contaminated with nonvolatile organic compounds, by causing reaction with bound indigenous organics (Zhu et al., 1992), and chromium, by causing conversion from the toxic hexavalent form to the nontoxic trivalent form (Sedhom et al., 1992). The feasibility of these processes has been demonstrated on a bench scale in a multimode oven. However, the challenge of bringing applicators and sufficient power to waste sites is formidable. If the promise of these applications is to be realized, additional work needs to be done to develop applicators for *in situ* processing and to show applicability and cost effectiveness on a larger scale in the field.

The cleanup of surface layers (0.5—5 cm depth) of concrete structures contaminated with radioactive or hazardous materials is a potentially costly problem affecting research laboratories, power plants, and processing and storage facilities. Mechanical removal methods create potentially hazardous dust, may drive contamination into the interior, or may create a large waste stream of contaminated water generated in dust amelioration efforts. Microwaves have been shown, in experiments in Japan (Yasunaka et al., 1987), Britain (Hills, 1989), and the United States (White et al., 1991b), to be effective in rapidly removing the outer layer of concrete in a dry process with reduced dust generation. Work is underway at Oak Ridge National Laboratory to scale-up the microwave concrete-removal process and to develop, build, and test a full-scale prototype. It is believed that removal rates exceeding those attainable through mechanical techniques are possible with optimized power, frequency, and applicator design. A schematic of the prototype apparatus is shown in Figure 5-7.

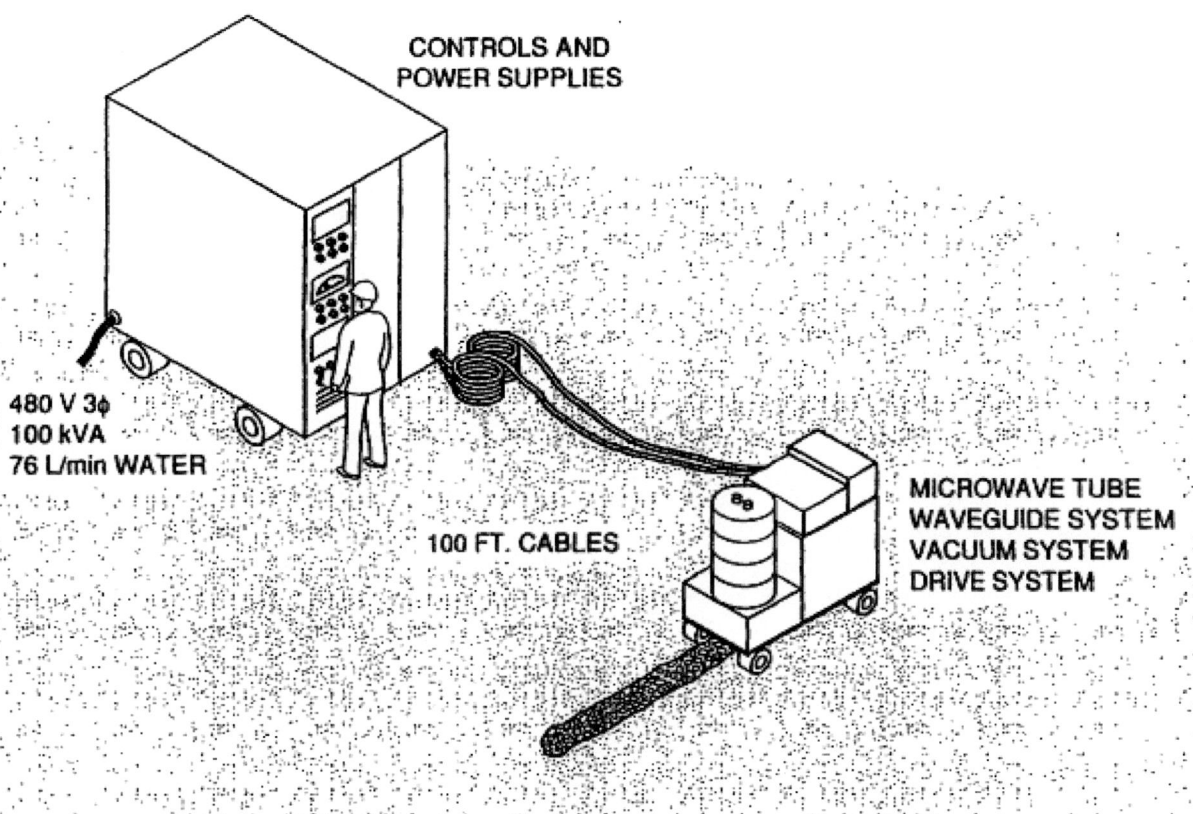

FIGURE 5-7
Schematic of prototype concrete-scabbing apparatus. (Courtesy of Dr. Terry L. White, Fusion Energy Division, Oak Ridge National Laboratory, Department of Energy).

SUMMARY

While a wide variety of materials have been processed using microwaves, including rubber, polymers, ceramics, composites, minerals, soils, wastes, chemicals, and powders, there are characteristics that make some materials very difficult to process. First, materials with significant ionic or metallic conductivity cannot be effectively processed due to inadequate penetration of the microwave energy. Second, insulators with low dielectric loss, including oxide ceramics and thermoplastic polymers, are difficult to heat from room temperature due to their low absorption of the incident energy. Since permittivity and loss factors often increase with temperature, hybrid heating may be used to process these types of materials by using alternate or indirect heating to raise the temperature of the parts to where they can be more effectively heated with microwaves. Finally, materials with permittivity or loss factors that increase rapidly during processing, such as alumina, can exhibit hot spots and thermal runaway. Although insulation or hybrid heating can improve the situation, stable microwave heating of these types of materials is problematic.

Enhanced apparent process kinetics due to microwave processing have been claimed for a range of materials, most notably ceramic sintering and polymer curing. However, in most cases, insufficient care was taken in temperature control and measurement and in measurement of critical process variables and material physical properties. A series of careful experiments with an internal calibration of the temperature is needed to eliminate the doubts that remain about the microwave enhancement effects.

Further investigation is needed to develop maps of the regimes of microwave-power absorption characteristics, batch size, heating rate, and other variables where microwave processing can be reproducible and uniform. This would allow processors to make informed decisions concerning microwave applications and process and equipment selection, while avoiding inefficient heating, uneven heating, and thermal runaway problems that have plagued earlier attempts.

Specific processes that show promise for future development include:

- ceramic processes including drying, chemical vapor infiltration, reaction bonding of silicon nitride, powder synthesis, and joining;
- polymeric composite pultrusion, ultradrawing of polymeric fibers, and adhesive bonding with intrinsically conducting organic polymers;
- chemical processes, including custom organic synthesis, hazardous materials processing, solvent extraction, and drying; and
- industrial waste processing, including treatment or remediation of process wastes, detoxification or consolidation of stored waste, and cleanup of storage or disposal sites.

In general, the elements required for successful application of microwave processing to industrial materials include selection of materials amenable to microwave processing; an understanding of the process requirements; an understanding of the process economics; characterization of material thermochemical properties; selection of equipment and design of applicators suitable for the application; an understanding of how the parts to be processed will interact with the microwave field; and adequate measurement and control of process variables such as incident power, part temperature, and field strength.

6

CONCLUSIONS AND RECOMMENDATIONS

Microwave processing of materials is a technology that can provide the material processor a new, powerful, and significantly different tool with which to process materials that may not be amenable to conventional means of processing or to improve the performance characteristics of existing materials. However, because it is fundamentally a new and different processing technology, it requires the materials processor to learn and understand the technology before attempting to use it.

The committee found that efforts in microwave process development that succeeded commercially did so because there was a compelling advantage for the use of microwave energy. Failure almost always resulted from simple, general causes, e.g., trying to process materials that were not conducive to microwave absorption or trying to use equipment that was not optimized for the particular material and application.

The most likely candidates for future production-scale applications will take full advantage of the unique characteristics of microwaves. For example, chemical vapor infiltration of ceramics and solution chemical reactions are enhanced by reverse thermal gradients that can be established using microwaves. Polymer, ceramic, and composite joining processes and catalytic processes are enabled by selective microwave heating. Powder synthesis of nanoparticles can take full advantage of rapid microwave heating to produce unique formulations and small particle sizes. Thermoplastic composite lamination and composite pultrusion processes are enhanced by rapid and bulk heating and by the ability to tailor the material's dielectric properties to microwave processes. The potential for portability and remote processing also make microwave processing attractive for waste remediation.

Due to the high cost of microwave generators and the relatively poor efficiency of electric power for heating applications, factors other than energy generally account for savings realized from microwave processing. Such factors include process time savings, increased process yield, and environmental compatibility.

The substantive general conclusions of the committee are summarized in this chapter. Recommendations to address technical shortfalls identified by the committee or to increase the probability of successful application of microwave processing technology are presented.

APPLICATIONS DEVELOPMENT

The future of microwave processing of materials appears to be strongest in specialty applications, and it will probably be of limited usefulness as a general method of producing process heat. Within the specialized areas, microwave processing has distinct advantages over conventional processing means. Microwave processing will not be applicable to all materials and in fact may be readily applicable only to certain types of materials.

The development of hybrid heating systems that optimally combine microwave sources with conventional sources to balance process variables such as required power, process flow time, tooling requirements, etc., represents a very promising, largely untapped area in process development. Hybrid heating may be provided actively, using a separate conventional heat source, or passively, using higher dielectric loss susceptors, insulation, or coatings that more readily absorb the incident power. Development of hybrid heating systems may be required for full realization of the benefits of microwave technology.

Most of the current research has focused on laboratory-scale, exploratory efforts. In order to realize the potential of microwave and hybrid processes, work is needed to scale-up process and system designs to large-batch or continuous processes. Process scaling includes model simulation, system design and integration, and an understanding of the costs and benefits involved in moving to production scale.

Recommendations:

- For particular materials, define the conditions under which microwaves provide uniform, stable processing. These may be developed through appropriate numerical modeling techniques and should be presented as processing charts that contain information on material properties, processing conditions, and specimen size and geometry. This modeling requires characterization of the thermal and physical properties of materials, including thermal conductivity and diffusivity, thermal expansion, and the temperature-dependent dielectric properties. Hybrid heating schemes, in which traditional heating is augmented with microwave heat, should be considered.
- Emphasize research work that facilitates the transition of developmental processes to production scale. This may include materials property characterization, process simulation, control schemes, equipment prototyping, and pilot-scale production.

PROCESS MODELING AND SIMULATION

An important element of microwave process development and system design is the capability to model electromagnetic interactions. An understanding of the variation of dielectric properties with temperature and processing state is crucial for simulations and process modeling. Computer modeling can be used to optimize generator or applicator system design, establish achievable processing windows, and conduct realistic process simulations for given dielectric properties, sample size, and desired processing conditions.

Recommendations:

- Compile existing material-property information on dielectric, magnetic, and thermal properties (including dependence on frequency and temperature) in the range useful in the processing of materials.
- Provide more-complete and more-consistent measurements of basic dielectric properties of materials to be processed using microwaves, and develop calibration standards for comparing the various techniques for dielectric properties measurements.
- Develop empirically simplified models and "microwave heating diagrams" based on measurements and on the extensive data collected from results of numerical simulation to make numerical techniques more accessible to processors.

SYSTEM DESIGN AND INTEGRATION

Failure to realize expected benefits from microwave processing is a result of inadequate interaction among researchers, materials engineers, process designers, and microwave engineers. In most cases, the basic equipment (e.g., generators, applicators, power supplies) for microwave processing applications is commercially available. However, the methodology for system integration, including system design, special applicator design, rapid equipment prototyping, and process control, is inadequate. It must be recognized that samples cannot be heated efficiently and uniformly if simply placed in a microwave oven without consideration of specific microwave/ materials interactions.

Recommendations:

- Establish multidisciplinary teams, consisting of materials and process engineers, microwave engineers, equipment designers, and manufacturing specialists, to properly develop microwave processes and procedures.
- Provide training in fundamentals of microwave processing technology, including microwave interactions with materials. Examples of available training opportunities include AFTER (Air Force Thermionic Engineering Research) and CAEME (Computer Applications in Electromagnetic Education) software for studying fundamental microwave interaction with materials.
- Define general specifications for applicator design, and characterize the resulting electromagnetic field to enable users to successfully apply microwaves to materials processing.
- Develop practical methods to monitor or determine internal temperature and thermal profiles (thermal gradients) within a material during the process cycle.

NONTHERMAL MICROWAVE EFFECTS

Although there is evidence of enhancements of processes due to the effects of microwaves alone (e.g., enhanced ceramic sintering, grain growth, and diffusion rates, and faster apparent kinetics in polymers and synthetic chemistry), the evidence is equivocal due to questionable temperature measurement techniques, uncertain process characterization methods, and conflicting evidence.

Recommendations:

- Establish standards for measurement of temperature to ensure reproducibility. In addition, the techniques and procedures used to measure temperature should be reported in detail, so an evaluation of accuracy can be made. The level of uncertainty in temperature measurements should also be reported. Perform experiments using several temperature-measurement techniques to determine the relative accuracy and reproducibility of the various techniques against a known standard (melting point, phase-transition temperature, etc., of well-characterized materials).
- Conduct detailed and controlled experiments to determine if microwave enhancement effects reported for materials are valid. Care should be taken to use a microwave source with predictable and reproducible fields and to have an internal temperature calibration to avoid temperature measurement uncertainties.

References

Agrawal, D. K., Y. Fang, D. M. Roy, and R. Roy. 1992. Fabrication of Hydroxyapatite Ceramics by Microwave Processing. Pp. 231—236 in Materials Research Society Symposium Proceedings, Vol. 269, Microwave Processing of Materials III. R. L. Beatty, W. H. Sutton, and M. F. Iskander, eds. Pittsburgh, Pennsylvania: Materials Research Society.

Agrawal, R. J. and L. T. Drzal. 1989. Effects of Microwave Processing on Fiber—Matrix Adhesion in Composites. Journal of Adhesion 29 (1-4):63—79.

Ahmad, I., R. Dalton, and D. Clark. 1991. Unique Application of Microwave Energy to the Processing of Ceramic Materials. Journal Microwave Power and Electromagnetic Energy. 26(3): 128—138.

Akyel, C. and E. Bilgen. 1989. Microwave and Radio-frequency Curing of Polymers: Energy Requirements, Cost and Market Penetration. Energy. 14(12):839—851.

Al-Assafi, S. and D. E. Clark. 1992. Microwave Joining of Ceramics: A Study in Joining Alumina Both Directly and with Alumina Gel. Pp. 335—340 in Materials Research Society Symposium Proceedings, Vol. 269, Microwave Processing of Materials III. R. L. Beatty, W. H. Sutton, and M. F. Iskander, eds. Pittsburgh, Pennsylvania: Materials Research Society.

Altschuler, H. 1963. Dielectric Constant. Handbook of Microwave Measurements, Vol 2. M. Sucher and J. Fox, eds. New York: Polytechnic Press.

Amano, M. and K. Nakagawa. 1987a. Drawing Behavior of Polymer Blends Consisting of Poly(ethyleneterephthalate) and Liquid Crystalline Copolyester. Polymer. 28(2):263—268.

Amano, M. and K. Nakagawa. 1987b. Microwave Heat-Drawing of Poly(vinyl alcohol) and Ethylene—Vinyl Alcohol Copolymers . Polymer Communications. 28(4): 119—120.

REFERENCES

Andrade, O. M., M. F. Iskander, and S. Bringhurst. 1992. High Temperature Broadband Dielectric Properties Measurement Techniques. Pp. 527—539 in Materials Research Society Symposium Proceedings, Vol. 269, Microwave Processing of Materials III. R. L. Beatty, W. H. Sutton, and M. F. Iskander, eds. Pittsburgh, Pennsylvania: Materials Research Society.

Asmussen, J. 1992. Microwave Applicator Theory for Single Mode / Multimode Processing of Materials. Proceedings of the American Chemical Society, Division of Polymeric Materials: Science and Engineering. 66:341-342.

Asmussen, J., H. H. Lin, B. Manring, and R. Fritz. 1987. Single-Mode or Controlled Multimode Microwave Cavity Applicators for Precision Material Processing. Review of Scientific Instruments. 58:1477—1486.

Athey, T. W., M. A. Stuchly, and S. S. Stuchly. 1982. Measurement of Radio Frequency Permittivity of Biological Tissues with an Open-ended Coaxial Line, Part I. IEEE Transactions on Microwave Theory and Techniques. MTT-30:82—86.

Baghurst, D. R. and D. M. P. Mingos. 1992a. Superheating Effects Associated with Microwave Dielectric Heating. Journal of the Chemical Society, Chemical Communications. 9:674—677.

Baghurst, D. R. and D. M. P. Mingos. 1992b. A New Reaction Vessel for Accelerated Syntheses Using Microwave Dielectric Super-Heating Effects. Journal of the Chemical Society, Dalton Transactions. (7):1151-1155.

Balebanov, A. V., B. A. Butylin, V. K. Zhivotov, V. I. Krokvenko, R. M. Matolich, S. S. Macheret, G. I. Novikov, B. V. Potapkin, V. D. Rusanov, A. A. Fridman, and V. T. Yavorskii. 1985. Dissociation of Hydrogen Sulfide in a Plasma. Doklady Physical Chemistry, Proceedings of the Academy of Sciences of the USSR, translated from Doklady Akademii Nauk SSSR. 283(3):657—660.

Barmatz, M. and H. W. Jackson. 1992. Steady State Temperature Profile in a Sphere Heated by Microwaves. Pp. 97—103 in Materials Research Society Symposium Proceedings, Vol. 269, Microwave Processing of Materials III. R. L. Beatty, W. H. Sutton, and M. F. Iskander, eds. Pittsburgh, Pennsylvania: Materials Research Society.

Bassen, H., W. Herman, and R. Hoss. 1977. EM Probe with Fiber Optic Telemetry System. Microwave Journal. April:35—39.

REFERENCES

Batt, J. A., R. Rukus and M. Gilden. 1992. General Purpose High Temperature Microwave Measurement of Electromagnetic Properties. Pp. 553—559 in Materials Research Society Symposium Proceedings, Vol. 269, Microwave Processing of Materials III. R. L. Beatty, W. H. Sutton, and M. F. Iskander, eds. Pittsburgh, Pennsylvania: Materials Research Society.

Baziard, Y. and A. Gourdenne. 1988a. Crosslinking under Microwaves (2.45 GHz) of Aluminum Powder—Epoxy Resin Composites-I. Electrical Power Dependence. European Polymer Journal. 24(9):873—880.

Baziard, Y. and A. Gourdenne. 1988b. Crosslinking under Microwaves (2.45 GHz) of Aluminum Powder—Epoxy Resin Composites-II. Aluminum Concentration Dependence. European Polymer Journal. 24(9):881—888.

Becker, Richard. 1964. Electromagnetic Fields and Interaction. New York: Blaidell Publishing Co.

Bennett, C. E. G., N. A. McKinnon, and L. S. Williams. 1968. Sintering in Gas Discharges. Nature. 217:1287—1288.

Berteaud, A. J. and J. C. Badot. 1976. High Temperature Microwave Heating in Refractory Materials. Journal Microwave Power. 11(4):315—320.

Bible, D. W., R. J. Lauf, and C. A. Everleigh. 1992. Multikilowatt Variable Frequency Microwave Furnace. Pp. 77—81 in Materials Research Society Symposium Proceedings, Vol. 269, Microwave Processing of Materials III. R. L. Beatty, W. H. Sutton, and M. F. Iskander, eds. Pittsburgh, Pennsylvania: Materials Research Society.

Blackham, D. 1992. Measurements of Complex Permittivity and Permeability Using Vector Network Analyzers. Proceedings of the American Chemical Society, Division of Polymeric Materials: Science and Engineering. 66:468—469.

Bloch, F. 1928. Zeitschrift fuer Physik. 52:555.

Boch, P., N. Lequeux and P. Piluso. 1992. Reaction Sintering of Ceramic Materials by Microwave Heating. Pp. 211—222 in Materials Research Society Symposium Proceedings, Vol. 269, Microwave Processing of Materials III. R. L. Beatty, W. H. Sutton, and M. F. Iskander, eds. Pittsburgh, Pennsylvania: Materials Research Society.

Boey, F., I. Gosling, and S. W. Lye. 1992. High-Pressure Microwave Curing Process for an Epoxy-Matrix/Glass-fibre Composite. Journal of Materials Processing Technology. 29(1-4):311—319.

Bond, G., R. B. Moyes, and D. A. Whan. 1992. Catalytic Reactions in a Microwave Field, in Congress Proceedings: First World Congress on Microwave Chemistry. Clifton, Virginia: International Microwave Power Institute.

Booske, J. H., R. F. Cooper, and I. Dobson. 1992. Mechanisms for Nonthermal Effects on Ionic Mobility during Microwave Processing of Crystalline Solids. Journal of Materials Research. 7(2):495—501.

Booske, J. H., R. F. Cooper, S. Freeman, and B. Meng. 1993. Studies of Microwave Field Effects on Ion Transport in Ionic Crystalline Solids. Annual Meeting, The American Ceramic Society, April 1993.

Bose, A. K., M. S. Manhas, B. K. Banik, and E. W. Robb. 1993. Microwave-Induced Organic Reaction Enhancement (MORE) Chemistry: Techniques for Rapid, Safe and Inexpensive Synthesis. Pp. A-1-1—A-1-17 in EPRI TR-102252, Proceedings: Microwave-Induced Reactions Workshop, M. Burka, R. D. Weaver, and J. Higgins, eds. Pleasant Hills, California: Electric Power Research Institute.

Bouazizi, A. and A. Gourdenne. 1988. Interactions between Carbon Black—Epoxy Resin Composites and Continuous Microwaves-I. Electrical Power Dependence of the Rate of Crosslinking of the Epoxy Matrix. European Polymer Journal. 24(9):889—893.

Brandon, J. R., J. Samuels, and W. R. Hodgkins. 1992. Microwave Sintering of Oxide Ceramics. Pp. 237—243 in Materials Research Society Symposium Proceedings, Vol. 269, Microwave Processing of Materials III. R. L. Beatty, W. H. Sutton, and M. F. Iskander, eds. Pittsburgh, Pennsylvania: Materials Research Society.

Bringhurst, S., O. M. Andrade, and M. F. Iskander. High Temperature Dielectric Properties Measurements of Ceramics. Pp. 561—568 in Materials Research Society Symposium Proceedings, Vol. 269, Microwave Processing of Materials III. R. L. Beatty, W. H. Sutton, and M. F. Iskander, eds. Pittsburgh, Pennsylvania: Materials Research Society.

Bringhurst, S., M. F. Iskander, and O. M. Andrade. 1993. New Metallized Ceramic Coaxial Probe for High-Temperature Broadband Dielectric Properties Measurements . Pp. 503—510 in Ceramic Transactions, Vol. 36, Microwaves: Theory and Application in Materials Processing II. D. E. Clark, W. R. Tinga and J. R. Laia, eds. Westerville, Ohio: American Ceramic Society.

Bullard, D. E. and Lynch, D. C. 1992. Processing Ores in a Microwave Induced 'Cold' Plasma. Pp 107—113 in Materials Research Society Symposium Proceedings, Vol. 269, Microwave Processing of Materials III. R. L. Beatty, W. H. Sutton, and M. F. Iskander, eds. Pittsburgh, Pennsylvania: Materials Research Society.

REFERENCES

Bur, A. J. 1985. Dielectric Properties of Polymers at Microwave Frequencies: A Review. Polymer. 26(July):963—977.

Burdette, E. C., F. L. Cain, and L Seals. 1980. In Vivo Measurement Technique for Determining Dielectric Properties at VHF Through Microwave Frequencies. IEEE Transactions on Microwave Theory and Techniques. MTT-28:414—427.

Busch, J. V. 1994. Personal communication from J. Busch, IBIS Associates to T. Munns, NMAB. January 14.

Chabinsky, I. J. 1983a. The Practice of Microwave Preheating Rubber Prep Stocks in Compression and Transfer Molding Operations. Rubber World . 188(1):34—36.

Chabinsky, I. J. 1983b. Practical Applications of Microwave Energy in the Rubber Industry. Elastomerics. 115(1):17—20.

Chabinsky, I.J. 1988. Applications of Microwave Energy: Past, Present and Future Brave New Worlds. Pp. 17—29 in Materials Research Society Symposium Proceedings, Vol. 124, Microwave Processing of Materials. W. H. Sutton, M. H. Brooks, and I. J. Chabinsky, eds. Pittsburgh, Pennsylvania: Materials Research Society.

Chabinsky, I. J. and E. E. Eves, III. 1986. The Application of Microwave Energy in Drying, Calcining, and Firing of Ceramics. InterCeram, 6 (Dec.):30—35.

Chan, L. and A. Gourdenne. 1992. Microwave Curing of Unsaturated Polyesters. Proceedings of the American Chemical Society, Division of Polymeric Materials: Science and Engineering. 66:382—383.

Chapman, B., M. F. Iskander, R. L. Smith, and O. M. Andrade. 1992. Simulation of Sintering Experiments in Single Mode Cavities. Pp. 53—59 in Materials Research Society Symposium Proceedings, Vol. 269, Microwave Processing of Materials III. R. L. Beatty, W. H. Sutton, and M. F. Iskander, eds. Pittsburgh, Pennsylvania: Materials Research Society.

Chaussecourte, P., J. F. Lamaudiere, and B. Maestrali. 1992. Electromagnetic Field Modeling of Loaded Microwave Cavity. Pp. 69—81 in Materials Research Society Symposium Proceedings, Vol. 269, Microwave Processing of Materials III. R. L. Beatty, W. H. Sutton, and M. F. Iskander, eds. Pittsburgh, Pennsylvania: Materials Research Society.

Chen, H. Y., M. F. Iskander, and J. E. Penner. 1991. Empirical Formula for Optical Absorption by Fractal Aerosol Agglomerates. Applied Optics. 30(12):1547—1551.

REFERENCES

Chen, M., J. E. McGrath, and T. C. Ward. 1989. Microwave Radiation Calorimetry of Thermoplastics. Proceedings of the American Chemical Society, Division of Polymeric Materials: Science and Engineering. 60:443—447.

Chen, M., M. A. Zumbrum, J. C. Hedrick, J. E. McGrath, and T. C. Ward. 1991. Electromagnetic Processing of Polymers: II. Quantitative Investigations of Microwave Processed Thermoplastics (Microwave Calorimetry), Pp. 431—440 in Materials Research Society Symposium Proceedings vol. 189, Microwave Processing of Materials II. W. B. Snyder, W. H. Sutton, M. F. Iskander, and D. L. Johnson, eds. Pittsburgh, Pennsylvania: Materials Research Society.

Cheng, J., J. Qui, J. Zhou, and N. Ye. 1992. Densification Kinetics of Alumina During Microwave Sintering. Pp. 323—328 in Materials Research Society Symposium Proceedings, Vol. 269, Microwave Processing of Materials III. R. L. Beatty, W. H. Sutton, and M. F. Iskander, eds. Pittsburgh, Pennsylvania: Materials Research Society.

Colomban, P. and J. C. Badot. 1979. Microwave Heating: Firing Method for Ceramics by 1990. L'industrie Ceramique. 725:101—107.

Colomban, P. and J. C. Badot. 1978. Elaboration of Anisotropic Superconducting Ceramics (Na^+ β-alumina) by Microwave Heating. Materials Research Bulletin. 13:135—139.

Copper, J. R. 1986. Fiber Optic Sensors-- High-Temperature Precision for Industry. Industrial Heating.

Cozzi, A. D., D. K. Jones, Z. Fathi, and D. E. Clark. 1991. Microstructural Evolution of $YBa_2Cu_3O_{7-x}$ Using Microwave Energy. Pp. 357—364 in Ceramic Transactions, Vol. 21, Microwaves: Theory and Application in Materials Processing. D. E. Clark, F. D. Gac, and W. H. Sutton, eds. Westerville, Ohio: American Ceramic Society.

CRC. 1986. Handbook of Biological Effects of Electromagnetic Fields. C. Polk and E. Postow, eds. Boca Raton, Florida: CRC.

Dalton, R. C. I. Ahmad, and D. E. Clark. 1990. Combustion Synthesis Using Microwave Energy. Ceramic Engineering Science Proceedings 11 (10): 1729—42.

Daniels, E. J., J. B. L. Harkness, and R. D. Doctor. 1992. Economics of Microwave Plasma Dissociation of H_2S. Presentation at American Institute of Chemical Engineers 1992 Spring National Meeting, March 29—April 2, 1992.

Dauerman, L. 1992. Microwave Treatment of Hazardous Wastes: Overview. Presentation to National Materials Advisory Board Committee on Microwave Processing of Materials: Science and Engineering An Emerging Industrial Technology. August, 1992.

REFERENCES

Dé, A., I. Ahmad, E. D. Whitney, and D. E. Clark. 1991a. Effect of Green Microstructure and Processing Variables on the Microwave Sintering of Alumina. Pp. 283—288 in Materials Research Society Symposium Proceedings, Vol. 189, Microwave Processing of Materials II. W. B. Snyder, W. H. Sutton, M. F. Iskander, and D. L. Johnson, eds. Pittsburgh, Pennsylvania: Materials Research Society.

Dé, A., I. Abroad, E. D. Whitney, and D. E. Clark. 1991b. Microwave (Hybrid) Heating of Alumina at 2.45 GHz: I. Microstructural Uniformity and Homogeneity. Pp. 319—328 in Ceramic Transactions, Vol. 21, Microwaves: Theory and Application in Materials Processing. D. E. Clark, F. D. Gac, and W. H. Sutton, eds. Westerville, Ohio: American Ceramic Society.

Dé, A., I. Abroad, E. D. Whitney, and D. E. Clark. 1991c. Microwave (Hybrid) Heating of Alumina at 2.45 GHz: II. Effect of Processing Variables, Heating Rates and Particle Size. Pp. 329—339 in Materials Research Society Symposium Proceedings, Vol. 189, Microwave Processing of Materials II. W. B. Snyder, W. H. Sutton, and M. F. Iskander, D. L. Johnson, eds. Pittsburgh, Pennsylvania: Materials Research Society.

Debye, P. 1929. Polar Molecules, Chemical Catalog Co., New York.

DeMeuse, M. T. 1992. Microwave Heating of PEEK. Proceedings of the American Chemical Society, Division of Polymeric Materials: Science and Engineering. 66:428—429.

Dils, R. R. 1983. High-Temperature Optical Fiber Thermometer. Journal Applied Physics. 54(3):1198—1201.

Dils, R. R., J. Geist, and M. L. Reilly. 1986. Measurement of the Silver Freezing Point with an Optical Fiber Thermometer: Proof of Concept. Journal Applied Physics. 59(4):1005—1012.

Drude, P. 1900. Annalen der Physik. 1: 566.

Drzal, L. T., K. J. Hook, and R. K. Agrawal. 1991. Enhanced Chemical Bonding at the Fiber—Matrix Interphase in Microwave Processed Composites. Pp. 449—454 in Materials Research Society Symposium Proceedings, Vol. 189, Microwave Processing of Materials II. W. B. Snyder, W. H. Sutton, M. F. Iskander, and D. L. Johnson, eds. Pittsburgh, Pennsylvania: Materials Research Society.

Dumey, C. H., M. F. Iskander, H. Massoudi, and C. C. Johnson. 1979. An Empirical Formula for Broad-Band SAR Calculations of Prolate Spheriodal Models of Humans and Animals. IEEE Transactions on Microwave Theory and Techniques. MTT-27:758—763.

REFERENCES

Eastman, J. A., K. E. Sickafus, J. D. Katz, S. G. Boeke, R. D. Blake, C. R. Evans, R. B. Schwarz, and Y. X. Liao. 1991. Microwave Sintering of Nanocrystalline TiO_2. Pp. 273—278 in Materials Research Society Symposium Proceedings, Vol. 189, Microwave Processing of Materials II. W. B. Snyder, W. H. Sutton, M. F. Iskander, and D. L. Johnson, eds. Pittsburgh, Pennsylvania: Materials Research Society.

Electric Power Research Institute. 1993. Proceedings: Microwave-Induced Reactions Workshop. EPRI TR-102252. M. Burka, R. D. Weaver, and J. Higgins, eds. Pleasant Hills, California: Electric Power Research Institute.

Epstein, A. J., J. Joo, C. Y. Wu, A. Benatar, C. F. Faisst, J. Zegarski, and A. G. MacDiarmid. 1993. Polyaniline: Recent Advances in Processing and Applications to Welding of Plastics. Pp. 165—178 in Proceedings NATO Advanced Research Workshop on Applications of Intrinsically Conducting Polymers. M. Aldissi, ed. Norwell, Massachusetts: Kluwer Academic Press .

Evans J. W., and D. Gupta. 1991. A Mathematical Model for Microwave-Assisted Chemical Vapor Infiltration. Pp. 101—107 in Materials Research Society Symposium Proceedings, Vol. 189, Microwave Processing of Materials II. W. B. Snyder, W. H. Sutton, M. F. Iskander, and D. L. Johnson, eds. Pittsburgh, Pennsylvania: Materials Research Society.

Fellows, L. A., S. Travis, and M. C. Hawley. 1993. Preliminary Comparison of Microwave Heating of Complex-Shaped Composites Reinforced with Conductive and Nonconductive Fibers. Pp. 415—424 in Materials Research Society Symposium Proceedings, Vol. 269, Microwave Processing of Materials III. R. L. Beatty, W. H. Sutton, and M. F. Iskander, eds. Pittsburgh, Pennsylvania: Materials Research Society.

Forrest, D. and J. Szekely. 1991. Global Warming the Primary Metals Industry. Journal of Metals. 43(12):23—30.

Frankel, J. 1926. Zeitschrift fuer Physik. 35:652.

Fukushima, H., T. Yamaka, and M. Matsui. 1990. Microwave Heating of Ceramics and its Application to Joining . Journal of Materials Research. 5(2):397—405.

George, C. E., I. Jun, and J. Fan. 1991. Application of Microwave Heating Techniques to the Detoxification of Contaminated Soils. Pp. 459—466 in Ceramic Transactions, Vol. 21, Microwaves: Theory and Application in Materials Processing. D. E. Clark, F. D. Gac, and W. H. Sutton, eds. Westerville, Ohio: American Ceramic Society.

Gourdenne, A. 1992. Interactions between Microwaves and Model Composite Materials. Proceedings of the American Chemical Society, Division of Polymeric Materials: Science and Engineering. 66:430—431.

REFERENCES

Haas, P. A. 1979. Heating of Uranium Oxides in a Microwave Oven. American Ceramic Society Bulletin. 58(9):873.

Harkness, J.B.L., A.J. Gorski and E.J. Daniels. 1990. Hydrogen Sulfide Waste Treatment by Microwave Plasma Dissociation. Presentation at 25th Intersociety Energy Conversion Engineering Conference (AIChE), August 1990.

Hawley, M. C. 1992. Features of Electromagnetic Processing of Polymers and Composites. Presentation to National Materials Advisory Board Committee on Microwave Processing of Materials: Science and Engineering An Emerging Industrial Technology. August, 1992.

Hedrick, J. C., D. A. Lewis, G. D. Lyle, S. D. Wu, T. C. Ward, and J. E. McGrath. 1989. Microwave Processing of Functionalized Poly (Arylene Ether Ketones). Proceedings of the American Chemical Society, Division of Polymeric Materials: Science and Engineering. 60: 438—442.

HHS. 1991. Food and Drug Administration 21 CFR 1030.10. U.S. Department of Health and Human Services. Public Health Service.

Hills, D. L. 1989. The Removal of Concrete Layers from Biological Shields by Microwaves. EUR 12185EN. Brussels: Commission of the European Communities.

Ho, W. W. 1988. High-Temperature Dielectric Properties of Polycrystalline Ceramics. Pp. 137—148 in Materials Research Society Symposium Proceedings, Vol. 124, Microwave Processing of Materials. W. H. Sutton, M. H. Brooks, and I. J. Chabinsky, eds. Pittsburgh, Pennsylvania: Materials Research Society.

Hollinger, R. D., V. K. Varadan and V. V. Varadan. 1993. Microwave Plasma Enhanced Deposition of Diamond Thin Films. Ceramic Transactions, Vol. 36. Theory and Application in Materials Processing II. D. E. Clark, F. D. Gac, and W. R. Tinga, eds. Westerville, Ohio: American Ceramic Society.

Holcombe, C. E., and N. L. Dykes. 1991a. "Ultra" High-Temperature Microwave Sintering. Pp. 375—385 in Ceramic Transactions, Vol. 21, Microwaves: Theory and Application in Materials Processing. D. E. Clark, F. D. Gac, and W. H. Sutton, eds. Westerville, Ohio: American Ceramic Society.

Holcombe, C. E. and N. L. Dykes. 1991b. Microwave Sintering of Titanium Diboride. Journal of Materials Science. 26(14):3730—3738.

International Microwave Power Institute. 1992. Congress Proceedings: First World Congress on Microwave Chemistry. Clifton, Virginia: International Microwave Power Institute.

REFERENCES

Ippen, J. 1971. Formulation for Continuous Vulcanization in Microwave Heating Systems. Rubber Chemistry and Technology. 44(1):294—306.

Iskander, M. F. 1993. Computer Modeling and Numerical Simulation of Microwave Heating Systems. Materials Research Society Bulletin. 18(11):30-36.

Iskander, M. F. 1992. Electromagnetic Fields and Waves. Englewood Cliffs, New Jersey: Prentice Hall.

Iskander, M. F. and J. B. DuBow. 1983. Time-and Frequency-domain Techniques for Measuring the Dielectric Properties of Rocks: A Review. Journal Microwave Power. 18(1):55—74.

Iskander, M. F., H. Massoudi, C. H. Durney, and S. J. Allen. 1981. Measurement of the RF Power Absorption in Spheroidal Human and Animal Phantoms Exposed to the Near Field of a Dipole Source. IEEE Transactions on Biomedical Engineering. BME-28:258—264.

Iskander, M. F., O. Andrade, A. Virkar, and H. Kimrey. 1991. Microwave Processing of Ceramics at the University of Utah-- Description of Activities and Summary of Progress. Pp. 35—48 in Ceramic Transactions, Vol. 21, Microwaves: Theory and Application in Materials Processing. D. E. Clark, F. D. Gac, and W. H. Sutton, eds. Westerville, Ohio: American Ceramic Society.

Iskander, M. F., R. L. Smith, O. Andrade, H. Kimrey, and L. Walsh. 1993. FDTD Simulation of Microwave Sintering of Ceramics in Multimode Cavities, IEEE Transactions on Microwave Theory and Techniques, accepted for publication.

Jackson, H. W. and M. Barmatz. 1991. Microwave Absorption by a Lossy Dielectric Sphere in a Rectangular Cavity. Journal of Applied Physics. 70(10):5193—5204.

Janney M. A. and H. D. Kimrey. 1992. Microwave Sintering of Solid Oxide Fuel Cell Materials, II: Lanthanum Chromite," submitted to Journal American Ceramic Society

Janney, M. A. and H. D. Kimrey. 1990. Microstructure Evolution in Microwave-Sintered Alumina. Ceramic Transactions. 7:382—390.

Janney, M. A. and H. D. Kimrey. 1988. Microwave Sintering of Alumina at 28 GHz. Pp. 919—924 in Ceramic Powder Science II. G. L. Messing, E. R. Fuller, and H. Hausner, Eds. Westerville, Ohio: American Ceramic Society.

REFERENCES

Janney, M. A., H. D. Kimrey, and J. O. Kiggans. 1992a. Microwave Processing of Ceramics: Guidelines Used at the Oak Ridge National Laboratory. Pp. 173—185 in Materials Research Society Symposium Proceedings, Vol. 269, Microwave Processing of Materials III. R. L. Beatty, W. H. Sutton, and M. F. Iskander, eds. Pittsburgh, Pennsylvania: Materials Research Society.

Janney, M. A., C. D. Calhoun, and H. D. Kimrey. 1992b. Microwave Sintering of Solid Oxide Fuel Cell Materials, I: Zirconia-8 mol% Yttria. Journal American Ceramic Society. 75(2):341—346.

Janney, M. A., H. D. Kimrey, M. A. Schmidt, and J. O. Kiggans. 1991a. Alumina Grain Growth in a Microwave Field. Journal American Ceramic Society 74(7):1675—1681.

Janney, M. A., C. L. Calhoun, and H. D. Kimrey. 1991b. Microwave Sintering of Zirconia -8 Mol % Yttria, Pp. 311—318 in Ceramic Transactions, Vol. 21, Microwaves: Theory and Application in Materials Processing. D. E. Clark, F. D. Gac, and W. H. Sutton, eds. Westerville, Ohio: American Ceramic Society.

Johnson, D. L. 1991. Microwave Processing of Ceramics at Northwestern University. Pp. 11—28 in Ceramic Transactions, Vol. 21, Microwaves: Theory and Application in Materials Processing. D. E. Clark, F. D. Gac, and W. H. Sutton, eds. Westerville, Ohio: American Ceramic Society.

Johnson, D. L., D. J. Skamser, and M. S. Spotz. 1993. Temperature Gradients in Microwave Processing: Boon and Bane. Pp. 133—146 in Ceramic Transactions, Vol. 36, Microwaves: Theory and Applications in Materials Processing II. D. E. Clark, W. R. Tinga, J. R. Saia, eds. Westerville, Ohio: American Ceramic Society.

Johnson, D. W. 1987. Innovations in Ceramic Powder Preparation. Pp. 3—19 in Advances in Ceramics, Vol. 21, Ceramic Powder Science. Messing, G. L., Mazdiyasni, K. S., McCauley, J. W. and Haber, R. A., eds. Westerville, Ohio: American Ceramic Society.

Jordan, C., J. Galy, J.P. Pascault, C. More, and C. Delmotte. 1992. A Comparison of a Standard Thermal Cure and a Microwave Cure of a Bulk Epoxy—Amine Matrix: Measurements of the Extent of Reaction and Mechanical Properties. Proceedings of the American Chemical Society, Division of Polymeric Materials: Science and Engineering. 66:380-381.

Jow, J., M. C. Hawley, M. Finzel, and T. Kern. 1988. Dielectric Analysis of Epoxy/Amine Resins Using Microwave Cavity Technique. Polymer Engineering and Science. 28(22):1450—1454.

Jow, J., J. D. DeLong, and M. C. Hawley. 1989. Computer-Controlled Pulsed Microwave Processing of Epoxy. SAMPE Quarterly. 20(2):46—50.

REFERENCES

Jow, J. 1988. Microwave Processing and Dielectric Diagnosis of Polymers and Composites Using a Single-Mode Resonant Cavity Technique. Ph.D. Thesis, Michigan State University.

Jullien, H. and A. Petit. 1992. The Microwave Reaction of Phenylglycidyl Ether with Aniline on Inorganic Supports as a Model for the Microwave Crosslinking of Epoxy Resins. Proceedings of the American Chemical Society, Division of Polymeric Materials: Science and Engineering. 66:378—379.

Katz, J. D. and R. D. Blake. 1991. Microwave Sintering of Multiple Alumina and Composite Components, Ceramic Bulletin. 70(8):1304—1308.

Katz, J.D., R. D. Blake, and V. M. Kenkre. 1991. Microwave Enhanced Diffusion?. Pp. 95—105 in Ceramic Transactions, Vol. 21, Microwaves: Theory and Application in Materials Processing. D. E. Clark, F. D. Gac, and W. H. Sutton, eds. Westerville, Ohio: American Ceramic Society.

Katz, J. D., R. D. Blake, J. J. Petrovic, and H. Sheinberg. 1988. Microwave Sintering of Boron Carbide. Pp. 219—226 in Materials Research Society Symposium Proceedings, Vol. 124, Microwave Processing of Materials. W. H. Sutton, M. H. Brooks, and I. J. Chabinsky, eds. Pittsburgh, Pennsylvania: Materials Research Society.

Kiggans, J. O., C. R. Hubbard, R. R. Steele, H. D. Kimrey, C. E. Holcombe, and T. N. Tiegs. 1991. Characterization of Silicon Nitride Synthesized by Microwave Heating. Pp. 403—410 in Ceramic Transactions, Vol. 21, Microwaves: Theory and Application in Materials Processing. D. E. Clark, F. D. Gac, and W. H. Sutton, eds. Westerville, Ohio: American Ceramic Society.

Kiggans, J. O. and T. N. Tiegs. 1992. Characterization of Sintered Reaction-Bonded Silicon Nitride Processed by Microwave Heating. Pp. 285—290 in Materials Research Society Symposium Proceedings, Vol. 269, Microwave Processing of Materials III. R. L. Beatty, W. H. Sutton, and M. F. Iskander, eds. Pittsburgh, Pennsylvania: Materials Research Society.

Kimrey, H. D. and M. F. Iskander. 1992. Microwave Processing of Dielectric Materials. Materials Research Society Short Course. April 25—26.

Kimrey, H. D., J. O. Kiggans, M. A. Janney, and R. L. Beatty. 1991. Microwave Sintering of Zirconia-Toughened Alumina Composites. Pp. 243—255 in Materials Research Society Symposium Proceedings, Vol. 189, Microwave Processing of Materials II. W. B. Snyder, W. H. Sutton, M. F. Iskander, and D. L. Johnson, eds. Pittsburgh, Pennsylvania: Materials Research Society.

REFERENCES

Kingston, H. M. 1992. Microwave Applications in Analytical Chemistry. In Congress Proceedings: First World Congress on Microwave Chemistry, Clifton, Virginia: International Microwave Power Institute.

Kingston, H. M. and L. B. Jassie. 1988a. Fundamental Relationships in Acid Decomposition of Samples for Elemental Analysis Using Microwave Energy. Pp. 121—128 in Materials Research Society Symposium Proceedings, Vol. 124, Microwave Processing of Materials. W. H. Sutton, M. H. Brooks, and I. J. Chabinsky, eds. Pittsburgh, Pennsylvania: Materials Research Society.

Kingston, H. M. and L. B. Jassie, eds. 1988b. Introduction to Microwave Sample Preparation: Theory and Practice. American Chemical Society Professional Reference Book. Washington: American Chemical Society.

Kittel, Charles. 1959. Solid State Physics, 2nd ed., New York: John Wiley and Sons.

Kladnig, W. E. and Horn, J.E. 1990. Submicron Oxide Powder Preparation by Microwave Processing. Ceramics International. 16(1):99—106.

Komarneni, S., R. Roy and Q. H. Li. 1992. Microwave-Hydrothermal Synthesis of Ceramic Powders. Materials Research Bulletin. 27:1393—1405.

Komarneni, S., Breval, E. and Roy, R. 1988. Microwave Preparation of Mullite Powders. Pp. 235—238 in Materials Research Society Symposium Proceedings, Vol. 124, Microwave Processing of Materials. W. H. Sutton, M. H. Brooks, and I. J. Chabinsky, eds. Pittsburgh, Pennsylvania: Materials Research Society.

Kozuka, H. and MacKenzie, J. D. 1991. Microwave Synthesis of Metal Carbides. Pp. 387—394 in Ceramic Transactions, Vol. 21, Microwaves: Theory and Application in Materials Processing. D. E. Clark, F. D. Gac, and W. H. Sutton, eds. Westerville, Ohio: American Ceramic Society.

Kraszewski, A. and S. S. Stuchly. 1983. Capacitance of Open-ended Dielectric-filled Coaxial Lines-- Experimental Results. IEEE Transactions on Instruments and Measurements. IM-32:517—519.

Kraszewski, A., M. A. Stuchly, and S. S. Stuchly. 1983. ANA Calibration Method for Measurements of Dielectric Properties. IEEE Transactions on Instruments and Measurements. IM-32:385—386.

Krage, M. K. 1981. Microwave Sintering of Ferrites. American Ceramic Society Bulletin. 60(11):1232—1234.

Krieger, B. 1992. Vulcanization of Rubber, A Resounding Success for Microwave Processing. Proceedings of the American Chemical Society, Division of Polymeric Materials: Science and Engineering. 66:339—340.

Krieger, B. 1989. Marketing Combination Heating Systems. Presented at Industrial Heating Workshop, International Microwave Power Institute. November 1989.

Kriegsmann, G. A. 1992. Thermal Runaway and its Control in Microwave Heated Ceramics. Pp. 257—264 in Materials Research Society Symposium Proceedings, Vol. 269, Microwave Processing of Materials III. R. L. Beatty, W. H. Sutton, and M. F. Iskander, eds. Pittsburgh, Pennsylvania: Materials Research Society.

Kumar, S. N., Pant, A., Sood, R. R., Ng-Yelim, H. and Holt, R. T. 1991. Production of Ultrafine Silicon Carbide by Fast Firing in Microwave and Resistive Furnaces . Pp. 391—402 in Ceramic Transactions, Vol. 21, Microwaves: Theory and Application in Materials Processing. D. E. Clark, F. D. Gac, and W. H. Sutton, eds. Westerville, Ohio: American Ceramic Society.

Largarov, A. N., S. M. Matitsin, and A. K. Sarychev. 1992. Microwave Properties of Polymer Materials Containing Conducting Inclusions. Proceedings of the American Chemical Society, Division of Polymeric Materials: Science and Engineering. 66:426—427.

Lauf, R. J., D. W. Bible, A. C. Johnson, and C. A. Everleigh. 1993. 2 to 18 GHz Broadband Microwave Heating Systems. Microwave Journal. 36(11):24—34.

Lauf, R. J., C. E. Holcombe, and C. Hamby. 1992. Microwave Sintering of Multilayer Ceramic Capacitors. Pp. 223—229 in Materials Research Society Symposium Proceedings, Vol. 269, Microwave Processing of Materials III. R. L. Beatty, W. H. Sutton, and M. F. Iskander, eds. Pittsburgh, Pennsylvania: Materials Research Society.

Lee, W. I. and G. Springer. 1984a. Interaction of Electromagnetic Radiation with Organic Matrix Composites . Journal of Composite Materials. 18:357—386.

Lee, W. I. and G. Springer. 1984b. Microwave Curing of Composites. Journal of Composite Materials. 18:387—409.

Levinson, L. M., H. A. Comanzo, and W. N. Schultz. 1992. Microwave Sintering of ZnO Varistor Ceramics. Pp. 311—321 in Materials Research Society Symposium Proceedings, Vol. 269, Microwave Processing of Materials III. R. L. Beatty, W. H. Sutton, and M. F. Iskander, eds. Pittsburgh, Pennsylvania: Materials Research Society.

REFERENCES

Lewis, D. A., J. D. Summers, T. C. Ward, and J. E. McGrath. 1992. Accelerated Imidization Reactions Using Microwave Radiation. Journal of Polymer Science: Polymer Chemistry. 30(8): 1647—1653.

Lewis, D. A., J. C. Hedrick, T. C. Ward, and J. E. McGrath. 1987. The Accelerated Curing of Epoxy Resins using Microwave Radiation. Polymer Preprints. 28(2):330—331.

Lewis, D. A., T. C. Ward, J. D. Summers, and J. E. McGrath. 1988. Cure Kinetics and Mechanical Behavior of Electromagnetically Processed Polyimides. Polymer Preprints. 29(1): 174—175.

Lind, A. C., L. N. Medgyesi-Mitshang, J. E. Kurz, H. F. McKinney, and F. C. Wear. 1991. Pp. 461—467 in Materials Research Society Symposium Proceedings, Vol. 189, Microwave Processing of Materials II. W. B. Snyder, W. H. Sutton, M. F. Iskander, and D. L. Johnson, eds. Pittsburgh, Pennsylvania: Materials Research Society.

Lorenson, C. and C. Gallerneault. 1991. Numerical Methods for the Modeling of Microwave Fields. Pp. 193—200 in Ceramic Transactions, Vol. 21, Microwaves: Theory and Application in Materials Processing. D. E. Clark, F. D. Gac, and W. H. Sutton, eds. Westerville, Ohio: American Ceramic Society.

Lorentz, H. A. 1904—1905. Amsterdam Proceedings.

Majetich, G. and R. Hicks. 1993. Applications of Microwave Accelerated Organic Chemistry. Pp. A-6-1—A-6-19 in EPRI TR-102252, Proceedings: Microwave-Induced Reactions Workshop, M. Burka, R. D. Weaver, and J. Higgins, eds. Pleasant Hills, California: Electric Power Research Institute.

Majetich, G. 1992. The Use of Commercial Microwave Ovens in Organic Chemistry! In Congress Proceedings: First World Congress on Microwave Chemistry, Clifton, Virginia: International Microwave Power Institute.

Manring, B. and J. Asmussen. 1991. Electromagnetic Modeling of Single-Mode Excited Material Loaded Applicators. Pp. 159—169 in Ceramic Transactions, Vol. 21, Microwaves: Theory and Application in Materials Processing. D. E. Clark, F. D. Gac, and W. H. Sutton, eds. Westerville, Ohio: American Ceramic Society.

Massachusetts Institute of Technology. 1948. Radiation Laboratory Series. New York: McGraw Hill.

McGee, T. D. 1988. Principles and Methods of Temperature Measurement. New York: John Wiley.

McMahon, G., A. Pant, R. Sood, A. Ahmad, and R. T. Holt. 1991. Microwave Sintering Technology for the Production of Metal Oxide Varistors. Pp. 237—242 in Materials Research Society Symposium Proceedings, Vol. 189, Microwave Processing of Materials II. W. B. Snyder, W. H. Sutton, M. F. Iskander, and D. L. Johnson, eds. Pittsburgh, Pennsylvania: Materials Research Society.

Meek, T. T., C. E. Holcombe and N. Dykes. 1987. Microwave Sintering of Some Oxide Materials Using Sintering Aids. Journal of Materials Science Letters. 6:1060—1062.

Mershon, J. Accurate High-Temperature Measurements in Microwave Environment. Luxtron (Accufiber) Technical Report.

Messing. G. L., Mazdiyasni, K. S., McCauley, J. W. and Haber, R. A., eds. 1987. Advances in Ceramics, Vol. 21, Ceramic Powder Science. Westerville, Ohio: American Ceramic Society.

Messing, G. L., Fuller, Jr., E. R., and Hausner, H., eds. 1988a. Ceramic Powder Science. Ceramic Transactions, American Ceramic Society Vol. 1, Part A. Westerville, Ohio: American Ceramic Society.

Messing, G. L., Fuller, Jr., E. R., and Hausner, H., eds. 1988b. Ceramic Powder Science. Ceramic Transactions, American Ceramic Society Vol. 1, Part B. Westerville, Ohio: American Ceramic Society.

Metaxas, A. C. and R. J. Meredith. 1983. Industrial Microwave Heating. Institute of Electrical Engineers. London: Peter Peregrinus, Ltd.

Methven, J. M. and S. R. Ghaffariyan. 1992. Microwave Assisted Pultrusion. Proceedings of the American Chemical Society, Division of Polymeric Materials: Science and Engineering. 66:389—390.

Michaelson, S. M. and J. C. Lin. 1987. Biological Effects and Health Implications of Radiofrequency Radiation. New York: Plenum Press.

Mijovic, J., A. Fishbain, and J. Wijaya. 1992a. Mechanistic Modeling of Epoxy-Amine Kinetics. 1. Model Compound Study. Macromolecules. 25(2):979—985.

Mijovic, J., A. Fishbain, and J. Wijaya. 1992b. Mechanistic Modeling of Epoxy-Amine Kinetics. 2. Comparison of Kinetics in Thermal and Microwave Fields. Macromolecules. 25(2):986—989.

Mijovic, J., and J. Wijaya. 1990. Comparative Calorimetric Study of Epoxy Cure by Microwave vs Thermal Energy. Macromolecules. 23 (15):3671—3674.

REFERENCES

Mijovic, J., J. Wijaya, and A. Fishbain. 1991. A Comparative Mechanistic Study of Epoxy/Amine Cure Kinetics in Thermal and Microwave Fields. Polymer Preprints. 32(3):366—367.

Mingos, D. M. P. and Baghurst, D. R. 1992. The Application of Microwaves to the Processing of Inorganic Materials. British Ceramic Society, Transactions and Journal. 91(4):124—127.

Mingos, D. M. P. 1993. The Applications of Microwaves in Chemical Synthesis. Pp. A-8-1—A-8-11 in EPRI TR-102252, Proceedings: Microwave-Induced Reactions Workshop, M. Burka, R. D. Weaver, and J. Higgins, eds. Pleasant Hills, California: Electric Power Research Institute.

Mingos, D. M. P. 1992. Microwave Dielectric Heating Effects in the Synthesis of Inorganic and Organometallic Compounds. In Congress Proceedings: First World Congress on Microwave Chemistry, Clifton, Virginia: International Microwave Power Institute.

Moisan, M. and J. Pelletier. 1992. Microwave Excited Plasmas. Pp. 435—497 in Plasma Technology, 4. Amsterdam: Elsevier.

Nakagawa, K., O. Maeda, and S. Yamakawa. 1983. Production of Ultrahigh Modulus Polyoxymethylene by Drawing Under Dielectric Heating. Journal Polymer Science: Polymer Letters. 21(11):933—935.

Nakagawa, K., T. Konaka, and S. Yamakawa. 1985. Production of Ultrahigh Modulus Polyoxymethylene by Drawing Under Dielectric Heating. Polymer. 26(1):84—88.

National Research Council. 1986. Advanced Processing of Electronic Materials in the United States and Japan. NMAB-443. Washington: National Academy Press.

National Research Council. 1990. Status and Applications of Diamond and Diamond-Like Materials: An Emerging Technology. NMAB-445. Washington: National Academy Press.

Navarro, A., M. J. Nunez, and E. Martin. 1991. Study of $TE_\circ$ and $TM_\circ$ Modes in Dielectric Resonators by a Finite Difference Time Domain Method Coupled with the Discrete Fourier Transform. IEEE Transactions on Microwave Theory and Techniques. 39:14—17.

Neas, E. 1992a. Microwave Chemistry: A Young Science with Great Potential. Presentation to the National Materials Advisory Board committee on Microwave Processing of Materials, December 1992.

Neas, E. 1992b. Basic Theoretical Considerations in Microwave Chemistry. In Congress Proceedings: First World Congress on Microwave Chemistry, Clifton, Virginia: International Microwave Power Institute.

Newnham, R. E., S. J. Jang. M. Xu, and F. Jones. 1991. Fundamental Interaction Mechanisms Between Microwaves and Matter. Ceramic Transactions, Vol. 21, Microwaves: Theory and Application in Materials Processing. D. E. Clark, F. D. Gac, and W. H. Sutton, eds. Westerville, Ohio: American Ceramic Society.

Nishitani, T. 1979. Method for Sintering Refractories and an Apparatus Therefor. U.S. Patent Number. 4147911, filed August 10, 1976, issued April 3, 1979.

Ochiai, T., M. Ikeda, and T. Nishitani. 1981. Application of Microwave Drying to Monolithic Ladle Linings. Taikabutsu Overseas. 1(2):92—96.

Oda, S. J. and I. S. Balbaa. 1988. Microwave Processing Research at Ontario Hydro Research Division. Pp. 303—309 in Materials Research Society Symposium Proceedings, Vol. 124, Microwave Processing of Materials. W. H. Sutton, M. H. Brooks, and I. J. Chabinsky, eds. Pittsburgh, Pennsylvania: Materials Research Society.

Oda. S. J., I. S. Balbaa, and B. T. Barber. 1991. The Development of New Microwave Heating Applications at Ontario Hydro's Research Division. Pp. 391—402 in Materials Research Society Symposium Proceedings, Vol. 189, Microwave Processing of Materials II. W. B. Snyder, W. H. Sutton, M. F. Iskander, and D. L. Johnson, eds. Pittsburgh, Pennsylvania: Materials Research Society.

Osepchuk, J. M. 1991. Radiofrequency Safety Issues in Industrial Heating Systems. Pp. 125—137 in Ceramic Transactions, Vol. 21, Microwaves: Theory and Application in Materials Processing. D. E. Clark, F. D. Gac, and W. H. Sutton, eds. Westerville, Ohio: American Ceramic Society.

OSHA. 1991. Occupational Safety and Health Administration 29 CFR 1910.97. U.S. Department of Labor.

Palaith, D., R. Silberglitt, C. M. Wu, R. Kleiner, and L. Libelo. 1988. Microwave Joining of Ceramics. Pp. 255—266 in Materials Research Society Symposium Proceedings, Vol. 124, Microwave Processing of Materials. W. H. Sutton, M. H. Brooks, and I. J. Chabinsky, eds. Pittsburgh, Pennsylvania: Materials Research Society.

Parker, T. G. 1972. Dielectric Properties of Polymers, II. Pp. 1297—1327 in Polymer Science, A. D. Jenkins, ed. Amsterdam: North-Holland Publishing.

REFERENCES

Patil, D. L., B. C. Mutsuddy, J. Gavulic, and M. Dahimene. 1991. Microwave Sintering of Alumina Ceramics in a Single Mode Applicator. Pp. 301—309 in Ceramic Transactions, Vol. 21, Microwaves: Theory and Application in Materials Processing. D. E. Clark, F. D. Gac, and W. H. Sutton, eds. Westerville, Ohio: American Ceramic Society.

Patterson, K. Y., Veillon and H. M. Kingston. 1988. Microwave Digestion of Biological Samples for Selenium Analysis by Electrothermal Atomic Assertion Spectrometry, in Introduction to Microwave Sample Preparation: Theory and Practice. American Chemical Society Professional Reference Book. Washington: American Chemical Society.

Patterson, M. C. L., P.S. Apte, R. M. Kimber, and R. Roy, 1992a. Batch Process for Microwave Sintering of Si_3N_4. Pp. 291—300 in Materials Research Society Symposium Proceedings, Vol. 269, Microwave Processing of Materials III. R. L. Beatty, W. H. Sutton, and M. F. Iskander, eds. Pittsburgh, Pennsylvania: Materials Research Society.

Patterson, M. C. L., P. S. Apte, R. M. Kimber, and R. Roy. 1992b. Mechanical and Physical Properties of Microwave Sintered Si_3N_4. Pp. 301—310 in Materials Research Society Symposium Proceedings, Vol. 269, Microwave Processing of Materials III. R. L. Beatty, W. H. Sutton, and M. F. Iskander, eds. Pittsburgh, Pennsylvania: Materials Research Society.

Patterson, M. C. L., R. M. Kimber, and P.S. Apté. 1991. The Properties of Alumina Sintered in a 2.45 GHz Microwave Field. Pp. 257—266 in Materials Research Society Symposium Proceedings, Vol. 189, Microwave Processing of Materials II. W. B. Snyder, W. H. Sutton, M. F. Iskander, and D. L. Johnson, eds. Pittsburgh, Pennsylvania: Materials Research Society.

Ragan, George L. 1948. Microwave Transmission Circuits. MIT Radiation Laboratory Series, Vol. 9. New York: McGraw-Hill.

Rains, R. C. 1988. Structural Adhesive Curing for Bonding Composites, Thermosets and Thermoplastics. Materials Research Society Symposium Proceedings, Vol. 124, Microwave Processing of Materials. W. H. Sutton, M. H. Brooks, and I. J. Chabinsky, eds. Pittsburgh, Pennsylvania: Materials Research Society.

Ramakrishna, D., S. Travis, and M. C. Hawley. 1993. Microwave Processing of Glass Fiber/Vinyl Ester—Vinyl Toluene Composites. Pp. 431—438 in Materials Research Society Symposium Proceedings, Vol. 269, Microwave Processing of Materials III . R. L. Beatty, W. H. Sutton, and M. F. Iskander, eds. Pittsburgh, Pennsylvania: Materials Research Society.

Ramo, S. and J. R. Whinnery. 1944. Fields and Waves in Modern Radio. New York: John Wiley & Sons.

REFERENCES

Redhead, C. S. 1992. Police Traffic Radar Safety. Congressional Research Service. Report 92-618 SPR.

Risman, P. 1991. Terminology and Notation of Microwave Power and Electromagnetic Energy. Journal Microwave Power and Electromagnetic Energy. 26(4):243—250.

Roussy, G., A. Mercier, J. Thiébaut and J. Vaugourg. 1985. Temperature Runaway of Microwave Heated Materials: Study and Control. Journal Microwave Power. 20(1):47—51.

Roussy, G., A. Bennani, and J. Thiebaut. 1987. Temperature Runaway of Microwave Irradiated Materials. Journal of Applied Physics. 62(4):1167—1170.

Roy, R., Komarneni, S. and Yang, L. J. 1985. Controlled Microwave Heating and Melting of Gels. Journal of the American Ceramic Society. 68(7):392—395.

Schroeder, R. E. and W. S. Hackett. 1971. Microwave Energy in the Foundry. British Foundryman. August: 293—298.

Schubring, N. W. 1983. Microwave Sintering of Alumina Spark Plug Insulators. Research Publication **GMR-4471**, General Motors Research Laboratories, Warren, Michigan.

Schwartz, H. F., R. G. Bosisio, M. R. Wertheimer, and D. Couderc. 1975. Microwave Curing of SR Compounds. Rubber Age. 107(10):27—38.

Sedhom, E., L. Dauerman, N. Ibrahim, and G. Windgasse. 1992. Microwave Treatment of Hazardous Wastes: "Fixation" of Chromium in Soil. Journal Microwave Power. 27(2):81—86.

Sheppard, L. M. 1988. Manufacturing Ceramics with Microwaves: The Potential for Economical Production. Ceramic Bulletin. 67(10):1556—1561.

Silberglitt, R., I. Ahmad, W. M. Black, and J. D. Katz. 1993. Recent Developments in Microwave Joining. Materials Research Society Bulletin. 18(11):47—50.

Simonian, J. L. 1979. Microwave Curing of Insulating Varnish on Armatures, Stators and Transformers. Proceedings of the 14th IEEE Electrical Insulation Conference. 105—109.

Singh, A. K., Mehta, P. and Kingon, A. I. 1991. Synthesis of Nonoxide Ceramic Powders by Nonthermal Microwave Plasma. Pp. 421—429 in Ceramic Transactions, Vol. 21, Microwaves: Theory and Application in Materials Processing. D. E. Clark, F. D. Gac, and W. H. Sutton, eds. Westerville, Ohio: American Ceramic Society.

REFERENCES

Skylarevich, V. E. and R. F. Decker. 1991. Super High Frequency Microwave Processing of Materials-A Basis for Developing New Technologies. Industrial Heating. October:54—56.

Skylarevich, V. E., A. Ketkov, M. Shevelev, and R. Decker. 1992. Interaction Between Gyrotron Radiation and Powder Materials. Pp. 163—169 in Materials Research Society Symposium Proceedings, Vol. 269, Microwave Processing of Materials III. R. L. Beatty, W. H. Sutton, and M. F. Iskander, eds. Pittsburgh, Pennsylvania: Materials Research Society.

Smith, F. J. 1991. Choosing the Right Electromagnetic Heat Process. Automation. 38(12):42—44.

Smith, F. J. 1988. Microwave Processing is Increasing, but it Needs Special Knowledge. Research and Development. 39(1):54—58.

Smith, R. D. 1991. Present and Future Uses of Microwave Power. Pp. 383—390 in Materials Research Society Symposium Proceedings, Vol 189, Microwave Processing of Materials II. W. B. Snyder, W. H. Sutton, M. F. Iskander, and D. L. Johnson, eds. Pittsburgh, Pennsylvania: Materials Research Society.

Smith, R. B. 1974. Industrial Applications of Microwave Energy. Transactions of International Microwave Power Institute. Vol. 2. Clifton, Virginia: International Microwave Power Institute.

Sommerfeld, A. 1928. Zeitschrift fuer Physik. 47:1.

Spotz, M. S., D. J. Skamser, P. S. Day, H. M. Jennings, and D. L. Johnson. 1993. Microwave-Assisted Chemical Vapor Infiltration. Ceramic Engineering and Science Proceedings. 14(9-10):753—760.

Spotz, M. S., D. J. Skamser, and D. L. Johnson. 1993. Thermal Stability of Dielectric Materials in Microwave Heating. Journal of the American Ceramic Society, accepted for publication.

Springer, G. S. 1992. Microwave Cure of Polymeric Matrix Composites, Microwave Processing of Polymers. Proceedings of the American Chemical Society; Division of Polymeric Materials: Science and Engineering. 66:420—421.

Stuchly, M. A. and S. S. Stuchly. 1980. Coaxial Line Reflection Methods for Measuring Dielectric Properties of Biological Substances at Radio and Microwave Frequencies-- A Review. IEEE Transactions on Instruments and Measurements. IM-29:176—183.

REFERENCES

Stuchly, M. A., T. W. Athey, and S. S. Stuchly. 1982. Measurement of Radio Frequency Permittivity of Biological Tissues with an Open-ended Coaxial Line, Part II. IEEE Transactions on Microwave Theory and Techniques. MTT-30:87—92.

Sutton, W. H. 1993. Key Issues in Microwave Process Technology. Pp. 3—18 in Ceramic Transactions, Vol. 36, Theory and Application in Materials Processing II. D. E. Clark, W. R. Tinga, and J. R. Laia, eds. Westerville, Ohio: American Ceramic Society.

Sutton, W. H. 1992. Microwave Processing of Ceramics-An Overview. Pp. 3—20 in Materials Research Society Symposium Proceedings, Vol. 269, Microwave Processing of Materials III. R. L. Beatty, W. H. Sutton, and M. F. Iskander, eds. Pittsburgh, Pennsylvania: Materials Research Society.

Sutton, W. H. 1989. Microwave Processing of Ceramics. American Ceramic Society Bulletin. 68(2):376—386.

Sutton, W. H. 1988. Microwave Firing of High Alumina Castables. Pp. 287—295 in Materials Research Society Symposium Proceedings, Vol. 124, Microwave Processing of Materials. W. H. Sutton, M. H. Brooks, and I. J. Chabinsky, eds. Pittsburgh, Pennsylvania: Materials Research Society.

Sutton, W. H. and W. E. Johnson. 1980. Method of Improving the Susceptibility of a Material to Microwave Energy Heating, U.S. Patent Number 4 219 361, filed June 9, 1978, issued August 26, 1980.

Sweeney, M. P. and D. L. Johnson. 1991. Microwave Plasma Sintering of Alumina. Pp. 365—372 in Ceramic Transactions, Vol. 21, Microwaves: Theory and Application in Materials Processing. D. E. Clark, F. D. Gac, and W. H. Sutton, eds. Westerville, Ohio: American Ceramic Society.

Takeuchi, Y., F. Yamamoto, K. Nakagawa, and S. Yamakawa. 1985. Orientation Behavior of High-Modulus Polyoxymethylene Produced by Microwave Heating Drawing. Journal of Polymer Science: Polymer Physics. 23(6):1193—1200.

Thomas, J. J., H. M. Jennings, and D. L. Johnson. 1993a. Microwave Nitridation of Silicon Compacts Utilizing a Temperature Gradient. Pp. 277—282 in Materials Research Society Symposium Proceedings, Vol. 287, Silicon Nitride Ceramics, Scientific and Technological Advances. I.-W. Chen, P. F. Becher, M. Mitomo, G. Petzow, and T.-S. Yen, eds. Pittsburgh, Pennsylvania: Materials Research Society.

Thomas, J. J., R. J. Christensen, D. L. Johnson, and H. M. Jennings. 1993b. Nonisothermal Microwave Processing of Reaction-Bonded Silicon Nitride. Journal of the American Ceramic Society 76(5):1384—1386.

REFERENCES

Thuery, J. 1992. Microwaves: Industrial, Scientific, and Medical Applications. Norwood, Massachusetts: Artech House.

Thuillier, F. M., H. Jullien, and M. F. Grenier-Loustalot. 1986. The Structure of Microwave-Cured Epoxy Resins Studied by FTIR and ^{13}C NMR CPMAS. Polymer Communications. 27(7):206—208.

Tian, Y-L. 1991. Practices of Ultra-Rapid Sintering of Ceramics Using Single Mode Applicators. Pp. 283—300 in Ceramic Transactions, Vol. 21, Microwaves: Theory and Application in Materials Processing. D. E. Clark, F. D. Gac, and W. H. Sutton, eds. Westerville, Ohio: American Ceramic Society.

Tian, Y. L., J. H. Feng, L. C. Sun, and C. J. Tu. 1992. Computer Modeling of Two Dimensional Temperature Distributions in Microwave Heated Ceramics. Pp. 41—46 in Materials Research Society Symposium Proceedings, Vol. 269, Microwave Processing of Materials III. R. L. Beatty, W. H. Sutton, and M. F. Iskander, eds. Pittsburgh, Pennsylvania: Materials Research Society.

Tian, Y. L., D. L. Johnson, and M. E. Brodwin. 1988a. Ultrafine Microstructure of Al_2O_3 Produced by Microwave Sintering. Pp. 925—932 in Ceramic Transactions Vol. I, Ceramic Powder Science 2, Part B. Westerville, Ohio: American Ceramic Society.

Tian, Y. L., D. L. Johnson, and M. E. Brodwin. 1988b. Microwave Sintering of Al_2O_3—TiC Composites. Pp. 933—938 in Ceramic Transactions, Vol. I, Ceramic Powder Science 2, Part B. Westerville, Ohio: American Ceramic Society.

Tiegs, T. N., J. O. Kiggans, and H. D. Kimrey Jr. 1991. Microwave Processing of Silicon Nitride. Pp. 267—272 in Materials Research Society Symposium Proceedings, Vol. 189, Microwave Processing of Materials II. W. B. Snyder, W. H. Sutton, M. F. Iskander, and D. L. Johnson, eds. Pittsburgh, Pennsylvania: Materials Research Society.

Tinga, W. R. 1992. Rapid High Temperature Measurement of Microwave Dielectric Properties. Pp. 505—516 in Materials Research Society Symposium Proceedings, Vol. 269, Microwave Processing of Materials III. R. L. Beatty, W. H. Sutton, and M. F. Iskander, eds. Pittsburgh, Pennsylvania: Materials Research Society.

Tinsley, F. G. B. and B. Adams. Evolution in the Application of Optical Fiber Thermometry. Accufiber Technical Note.

Valentine, J. M. 1973. Role of Microwaves in the Core-room. Foundry. October: 37—39.

Valentine, J. M. 1977. Production of Plaster Molds by Microwave Treatment. U. S. Patent 4 043 380. August 23.

REFERENCES

Van, Q. L. and A. Gourdenne. 1987. Microwave Curing of Epoxy Resins with Diaminodiphyenylmethane — I. General Features. European Polymer Journal. 23(10):777—780.

Varadan, V. K., H. S. Dewan, V. V. Varadan, and A. Lakhtakia. 1990. Microwave Joining of Polymers and Polymer Composites. Pp. 1766—1768 in ANTEC 90, Plastics in the Environment: Yesterday, Today, and Tomorrow. Brookfield, Connecticut: Society of Plastics Engineers.

Varma, R. and S. P. Nandi. 1991. Oxidative Degradation of Trichloroethylene Adsorbed on Active Carbons: Use of Microwave Energy. Pp. 467—473 in Ceramic Transactions, Vol. 21, Microwaves: Theory and Application in Materials Processing. D. E. Clark, F. D. Gac, and W. H. Sutton, eds. Westerville, Ohio: American Ceramic Society.

Varma, R., S. P. Nandi, and J. D. Katz. 1991. Detoxification of Hazardous Waste Streams Using Microwave-Assisted Fluid-Bed Oxidation. Pp. 67—68 in Materials Research Society Symposium Proceedings, Vol. 189, Microwave Processing of Materials II. W. B. Snyder, W. H. Sutton, M. F. Iskander, and D. L. Johnson, eds. Pittsburgh, Pennsylvania: Materials Research Society.

Vollath, D., Varma, R. and Sickafus, K. E. 1992. Synthesis of Nanocrystalline Powders for Oxide Ceramics by Microwave Plasma Pyrolysis. Pp. 379—384 in Materials Research Society Symposium Proceedings, Vol. 269, Microwave Processing of Materials III. R. L. Beatty, W. H. Sutton, and M. F. Iskander, eds. Pittsburgh, Pennsylvania: Materials Research Society.

Von Hippel, A. 1954. Dielectric Materials and Applications. New York: Technology Press of MIT and John Wiley & Sons.

Voss, W. A. G., and H. K. Kua. 1991. Model-Informed Microwave Processing of Materials. Pp. 3—14 in Materials Research Society Symposium Proceedings, Vol. 189, Microwave Processing of Materials II. W. B. Snyder, W. H. Sutton, M. F. Iskander, and D. L. Johnson, eds. Pittsburgh, Pennsylvania: Materials Research Society.

Wagner, C., and W. Schottky. 1930. Zeitschrift fuer Physikalische Chemie. B11:163.

Walkiewicz, J. W., A. E. Clark, and S. McGill. 1991. Microwave Assisted Grinding. IEEE Transactions on Industrial Applications. 27(2):239—243.

Walkiewicz, J. W., G. Kazonich, and S. L. McGill. 1988. Microwave Heating Characteristics of Selected Mineral Ores Materials and Compounds. Minerals and Metallurgical Processing. 5(1):39—42.

REFERENCES

Wan, J. K. S. and T. A. Koch. 1993. Application of Microwave Radiation for the Synthesis of Hydrogen Cyanide. Pp. A-3-1—A-3-13 in EPRI TR-102252, Proceedings: Microwave-Induced Reactions Workshop, M. Burka, R. D. Weaver, and J. Higgins, eds. Pleasant Hills, California: Electric Power Research Institute.

Ward, T. C. and M. Chen. 1992. Basic Ideas of Microwave Processing of Polymers. Proceedings of the American Chemical Society, Division of Polymeric Materials: Science and Engineering. 66:335—336.

Wei, J., M. C. Hawley, J. Jow, and J. D. Long. 1991. Microwave Processing of Crossply Continuous Graphite Fiber/Epoxy Composites. SAMPE Journal. 27(1):33—39.

Weil, C. M. 1992. The Electromagnetic Materials Program at NIST. Pp. 517-526 in Materials Research Society Symposium Proceedings, Vol. 269, Microwave Processing of Materials III. R. L. Beatty, W. H. Sutton, and M. F. Iskander, eds. Pittsburgh, Pennsylvania: Materials Research Society.

Wertheimer, M. R. and H. P. Schreiber. 1981. Surface Property Modification of Aromatic Polyamides by Microwave Plasmas. Journal of Applied Polymer Science. 26:2087—2096.

Westphal, W. B. and J. Iglesias. 1971. Dielectric Spectroscopy of High-Temperature Materials. Technical Report AFML-TR-71-66. Wright-Patterson Air Force Base, Ohio: Air Force Materials Laboratory.

Westphal, W. B. and A. Sils. 1972. Dielectric Constant and Loss Data. Technical Report AFML-TR-72-39. Wright-Patterson Air Force Base, Ohio: Air Force Materials Laboratory.

Westphal, W. B. 1975. Dielectric Constant and Loss Data, Part II. AFML-TR-74-250. Wright-Patterson Air Force Base, Ohio: Air Force Materials Laboratory.

Westphal, W. B. 1977. Dielectric Constant and Loss Data, Part III. Wright-Patterson Air Force Base, Ohio: Air Force Materials Laboratory.

Westphal, W. B. 1980. Dielectric Constant and Loss Data, Part IV. Wright-Patterson Air Force Base, Ohio: Air Force Materials Laboratory.

White, T. L., E. L. Youngblood, J. B. Berry, and A. J. Mattus. 1991a. First Results of In-can Microwave Processing Experiments for Radioactive Liquid Wastes at the Oak Ridge National Laboratory. Pp. 91—97 in Materials Research Society Symposium Proceedings, Vol. 189, Microwave Processing of Materials II. W. B. Snyder, W. H. Sutton, M. F. Iskander, and D. L. Johnson, eds. Pittsburgh, Pennsylvania: Materials Research Society.

REFERENCES

White, T. L., R. G. Grubb, L. P. Pugh, D. Foster, Jr., and W. D. Box. 1991b. Removal of Contaminated Concrete Surfaces by Microwave Heating-Phase I Results. Contract DE-AC05-84OR21400. Oak Ridge, Tennessee: Martin Marietta Energy Systems.

Willert-Porada, M. 1993. Microwave Processing of Metallorganics to Form Powders, Compacts, and Functional Gradient Materials. Materials Research Society Bulletin. 18(11):51—57.

Willert-Porada, M., T. Krummel, B. Rohde, and D. Moorman. 1992. Ceramic Powders by Metallorganic and Microwave Processing. Pp. 199—204 in Materials Research Society Symposium Proceedings, Vol. 269, Microwave Processing of Materials III. R. L. Beatty, W. H. Sutton, and M. F. Iskander, eds. Pittsburgh, Pennsylvania: Materials Research Society.

Windgasse, G. and L. Dauerman. 1992. Microwave Treatment of Hazardous Wastes: Removal of Volatile and Semi-Volatile Organic Contaminants From Soil. Journal Microwave Power. 27(1):23—32.

Wingard, C. D. and C. L. Beatty. 1990. Crosslinking of an Epoxy with a Mixed Amine as a Function of Stoichiometry. I. Cure Kinetics via Dynamic Mechanical Spectroscopy. Journal Applied Polymer Science. 40:1981—2005.

Woo, E. M. and J. C. Seferis. 1990. Cure Kinetics of Epoxy/Anhydride Thermosetting Matrix System. Journal Applied Polymer Science. 40:1237—1256.

Wu, C. Y., and A. Benatar. 1992. Microwave Joining of HDPE Using Conductive Polyaniline Composites. Proceedings Society of Plastics Engineers. 50th Annual Technical Conference. 1771—1774.

Yasunaka, H., M. Shibamoto, T. Sukegawa, T. Yamate, and T. Tanaka. 1987. Microwave Decontaminator for Concrete Surface Decontamination in JPDR. Pp IV/109—IV/115 in Proceedings of the International Decommissioning Symposium. Report CONF-871018, Vol. 2, G. A. Tarcza, ed. Springfield, Virginia: NTIS.

Yiin, T. Y., V. V. Varadan, V. K. Varadan, and J. C. Conway. 1991. Microwave Joining of Si-SiC/Al/Si-SiC. Pp. 507—514 in Ceramic Transactions, Vol. 21, Microwaves: Theory and Application in Materials Processing. D. E. Clark, F. D. Gac, and W. H. Sutton, eds. Westerville, Ohio: American Ceramic Society.

Yu, X. D., V. V. Varadan, V. K. Varadan. 1991. Application of Microwave Processing to Simultaneous Sintering and Joining of Ceramics. Pp. 497—503 in Ceramic Transactions, Vol. 21, Microwaves: Theory and Application in Materials Processing. D. E. Clark, F. D. Gac, and W. H. Sutton, eds. Westerville, Ohio: American Ceramic Society.

Zhang, J., L. Cao, and F. Xia. 1992. Microwave Sintering of Si_3N_4 Ceramics. Pp. 329—334 in Materials Research Society Symposium Proceedings, Vol. 269, Microwave Processing of Materials III. R. L. Beatty, W. H. Sutton, and M. F. Iskander, eds. Pittsburgh, Pennsylvania: Materials Research Society.

Zhu, N., L. Dauerman, H. Gu, and G. Windgasse. 1992. Microwave Treatment of Hazardous Wastes: Remediation of Soils Contaminated by Non-Volatile Organic Chemicals Like Dioxins. Journal Microwave Power. 27(1):54—61.

REFERENCES

APPENDIX: BIOGRAPHICAL SKETCHES OF COMMITTEE MEMBERS

DALE F. STEIN is professor of metallurgical and materials engineering and president emeritus of Michigan Technological University. Professor Stein received a B.S. from the University of Minnesota and a Ph.D. in metallurgy from Rensselaer Polytechnic Institute. He has expertise in a number of technical areas, including metallurgy, chemistry, and mechanical engineering. Professor Stein is a member of National Academy of Engineering.

RICHARD H. EDGAR is general manager of the Industrial Microwave Department at Amana Refrigeration, Inc. Mr. Edgar received a B.S. in electrical engineering from Northeastern University. His experience is in the design and commercial application of microwave tubes and equipment in industrial, medical, and military product lines. He is a member and past president of the International Microwave Power Institute.

MAGDY F. ISKANDER is professor of electrical engineering at the University of Utah and director of the Center for Computer Applications in Electromagnetic Education. Professor Iskander received a B.S. from the University of Alexandria (Egypt) and an M.S. and Ph.D. from the University of Manitoba (Canada), all in electrical engineering. His research and experience is in electromagnetic theory, microwave applications, computer process simulation, and the application of computers in education.

SYLVIA M. JOHNSON is program manager for ceramics at SRI International. Dr. Johnson received a B.S. from the University of New South Wales (Australia) and an M.S. and Ph.D. in engineering and materials science from the University of California, Berkeley. Her background is in oxide and nonoxide powder synthesis, processing of ceramics, and ceramic joining.

D. LYNN JOHNSON is professor of materials science at Northwestern University. Professor Johnson received a B.S. and Ph.D. in Ceramic Engineering from the University of Utah. His research is in ceramic sintering, transport properties, plasma and microwave processing, and processing of high-temperature superconductors.

CHESTER G. LOB is vice president and chief engineer at Varian Associates, Inc. Dr. Lob received a B.S. from Tulane University and an M.S. and Ph.D. in electrical engineering from the University of Illinois. His experience has been in the development and application of microwave tubes and devices and in microwave signal processing.

JANE M. SHAW is senior manager of Thin Film Materials and Processes at IBM's T. J. Watson Research Center. She received a B.S. in biology and chemistry from Elms College. Her experience has been in new fabrication techniques, polymers, metallization and interconnection technology for chip and packaging applications.

WILLARD H. SUTTON is senior research scientist at the United Technologies Research Center. Dr. Sutton received a B.S. from Alfred University and an M.S. and Ph.D. in ceramics technology from Pennsylvania State University. He has experience in ceramic technology as it is applied to metallurgical processes, especially in superalloy processing, vacuum melting and refining, high-temperature melt purification, and microwave firing of ceramic materials.

PING K. TIEN is fellow emeritus in the Photonics Research Laboratory at AT&T's Bell Laboratories. Dr. Tien received a B.S. from National Central University (China) and an M.S. and Ph.D. from Stanford University. His research is in device physics, microwave electronics, wave propagation, acoustics in solids, gas lasers, superconductivity, and integrated optics. He is a member of the National Academy of Sciences and the National Academy of Engineering.